Institut International de Philosophie
International Institute of Philosophy

Problèmes Philosophiques d'Aujourd'hui

Tome 3
Monde et Mondanéité

par les soins de
PETER KEMP
Université danoise de l'Éducation,
Copenhague, Le Danemark.

Springer

Philosophical Problems Today

Volume 3
World and Worldhood

edited by
PETER KEMP
Danish University of Education,
Copenhagen, Denmark

Springer

A C.I.P. Catalogue record for this book is available from the Library of Congress.

ISBN 978-90-481-6769-2 (PB)
ISBN 978-1-4020-3027-7 (e-Book)

Published by Springer,
P.O. Box 17, 3300 AA Dordrecht, The Netherlands.

Sold and distributed in North, Central and South America
by Springer,
101 Philip Drive, Norwell, MA 02061, U.S.A.

In all other countries, sold and distributed
by Springer,
P.O. Box 322, 3300 AH Dordrecht, The Netherlands.

Published under the auspices of the International Council of Philosophy and Humanistic Studies and of the International Federation of Philosophical Societies, with the support of UNESCO.

Publié sous les auspices du Conseil International de la Philosophie et des Sciences Humaines et de la Fédération Internationale des Sociétés de Philosophie, avec le concours de l'UNESCO.

Printed on acid-free paper

springeronline.com

TABLE OF CONTENTS / TABLE DES MATIÈRES

PREFACE

The idea of globalization (*mondialisation*) is in fashion. But what, in this globalization, does « globe » or « world » (*monde*) signify ? This book will attempt to answer this question, first by analysing the origin and development of the concept of world from Greek thought up to the 20th century, and then by examining perspectives on the world, insofar as it is becoming a technological, economic, aesthetic, and political world.

Rémi Brague returns to the Greek idea of the *cosmos* in order to track the worldhood (*mondanité*) of the world, that is, the process by which the idea of the world is formed. He shows that this idea was never intended to be a description of a reality – of, for instance, physical reality – but always expressed a value judgement. If the universe constitutes the world, it is because it is an order of things, a celestial order, that humans are called to imitate in their moral lives. But this world is not an external authority we are required to obey. The *cosmos* is in us in the same way as, for Kant, the moral law lives in us.

But this conception of the world has disappeared today. Brague states that cosmology as moral guide no longer exists ; physical nature no longer speaks to us, and it can even appear terrifying, as witnessed in the 18th century with the earthquake of Lisbon. On the contrary, we have developed « worldviews », or *Weltanschauungen*, that express the value of non-physical realities. The world in which we can henceforth live is a human world – that is, a social, moral, or « intelligible » world. The world is now a way of seeing and living reality, and things and people attain their membership in the world through a process of « worlding / world-constitution » (*mondanéité*). We now constitute our world or our own particular worlds, for to exist as human reality is to « be-in-the-world », as Heidegger says.

It is in this way that the world has become a world of technologically produced human works. Don Ihde shows how a world has developed in which technologies are no longer considered neutral means serving the ends of human action, but become the very means by which people exist in the world. From this perspective he envisions three main areas of difficulty for a philosophy of technology : first, how to react with respect to the system of interactions that constitutes our environment, in order to rehabilitate a polluted world and avoid the major accidents instigated by high technology and big industry ; second, how to resolve ethnic conflicts and promote respect for values of tolerance and sympathy in ever more multicultural societies ; and finally, how to control the biotechnological manipulation of the microscopic world. All these problems attest to the fact that people are no longer neutral observers of the world, but more or less informed participants.

This participation also has an economic dimension. Vittorio Mathieu construes economics as the science of the role of money at two levels – that of the individual and that of society. In trade between individuals each can pursue his own interests, but for society the result of commerce is something other than the ends pursued by each economic agent – that is, it is the distribution of goods and services not expressly intended by the particular purchases and sales of merchants. One can carry out, however, non-economic – that is to say political – interventions in economic life. But the effects of such interventions are limited, for no economy can exist without the freedom of each to transform his money into the production or acquisition of products or services. To possess money is to have freedom of choice ; it is to have a power that permits one to wait, and thus to have time at one's disposal. And if the value of money decreases through inflation, this is certainly an economic fact, but its cause is not economic. Inflation is the result of a form of robbery, a political evil that must be combatted. Citizens, in a state characterized by material abundance, on the other hand, would be assured of a basic income in order that they be able to freely dispose of their own time.

But the economic world is not the only sphere of freedom. Tomonobu Imamichi, an eco-ethical philosopher pursuing the conjunction of technology and the good life, analyses the capacity of aesthetic experience to disclose a world other than those worlds that arise from the destruction of the « fullness of the moment » through the technology of time-acceleration. Through the experience of art, people can resist two opposing worldviews, neither of which leaves room for human freedom : that of a mechanical-mathematical world, and that of a world of animal unconsciousness.

In the eastern tradition this aesthetic experience is tied to a religious experience, which makes possible the ecstasy that enables the human spirit to distance itself from the world of technological efficiency. Thus for

Imamichi art is the opening toward « the horizon of eternal being that technology tries to supersede with its continuous functioning ».

It must also not be forgotten that politics creates a world in which great tasks are presented to us. On the basis of a conception of the world of politics as a world of power, Francisco Miró Quesada C. emphasises that the great political questions are not solvable without worldviews that express value systems. A purely pragmatic worldview is not sufficient, for example, to enable one to respect human rights and oppose war, racism, etc. ; also needed is a philosophical reflection that subscribes to a rational ideal of life.

David Rasmussen returns to this ideal in Kant's view of history « from the cosmopolitan point of view », and in the idea of reasonability in John Rawls' *The Law of Peoples*. Rasmussen thinks that to prescribe not only a rational but a reasonable coexistence among so-called liberal and non-liberal societies, as Rawls does, implies a weak cosmopolitanism, even though Rawls doesn't see himself as cosmopolitan. What is in fact at issue is a view of the world based on the aesthetic judgment that humans share in being sensible beings, and which thus allows one to adopt fundamental common principles of practical and political life and therein to fulfill the obligation to live together. In this way, *sensus communis* appears as a cosmopolitan concept, which founds a political globalization of the world.

Thus, this book inscribes itself into the current debate on globalization, which integrates, for good and bad, all expressions of human life into one global system. The danger of this global transformation of the world arises from the fact that all local, regional, and national power risks being neutralised for the benefit of an anonymous and indeterminable techno-economic power. In what follows, I attempt to grasp the meaning of that globalization upon which the destiny of our planet depends.

Our thanks are due to the Centre national de la recherche scientifique (Paris), to the Conseil international de la philosophie et des sciences humaines (UNESCO), and to the staff at Kluwer Academic Publishers (Dordrecht).

Peter KEMP.

PRÉFACE

L'idée de mondialisation est à la mode. Mais qu'est-ce que signifie « monde » dans cette mondialisation ? C'est une question philosophique à laquelle nous cherchons à répondre dans cet ouvrage en analysant l'origine et le développement du concept même dans la pensée grecque jusqu'au 20e siècle, puis en examinant les perspectives du monde dans la mesure où il devient un monde technologique, économique, esthétique et politique.

Rémi Brague retourne à l'idée grecque du *cosmos* pour suivre « *la mondanité* » du monde, c'est-à-dire le processus par lequel se forme l'idée du monde. Il montre que cette idée n'a jamais voulu être une description d'une réalité, par exemple la réalité physique, mais a toujours exprimé un jugement de valeur. Si l'univers constitue le monde, c'est parce qu'il est un ordre des choses, un ordre céleste, que l'homme est appelé à imiter dans sa vie morale. Mais ce monde n'est pas une instance extérieure à laquelle il s'agirait d'obéir. Le *cosmos* est en nous de la même manière que la loi morale chez Kant habite notre intérieur.

Or cette conception du monde a disparu aujourd'hui. Brague constate qu'il n'y a plus de cosmologie comme guide moral ; la nature physique ne nous parle plus, elle peut même se montrer terrifiante, comme en témoignait au 18e siècle le tremblement de terre de Lisbonne. Par contre, nous avons développé des « visions du monde », des *Weltanschauungen,* qui expriment la valeur des réalités non-physiques. Le monde dans lequel on peut vivre est désormais un monde humain, que ce soit le monde social, moral, le « monde intelligible » ou autre. Le monde est maintenant une manière de voir et de vivre la réalité, et les choses et les hommes acquièrent leur appartenance à un monde par un processus de *mondanéité*. Nous constituons maintenant notre monde ou nos propres mondes, car exister comme réalité humaine est « être-dans-le-monde », comme le dit Heidegger.

C'est ainsi que le monde est devenu un monde des œuvres humaines créées par la technologie. Don Ihde montre comment s'est fait un monde où les technologies ne sont plus considérées comme des moyens neutres en vue des fins de l'action humaine, mais deviennent des moyens par lesquels l'homme existe dans le monde. Dans cette perspective il voit trois grands domaines de problèmes pour une philosophie de la technologie : d'abord comment réagir par rapport au système d'interactions qui constitue notre environnement pour rétablir un monde non-pollué et éviter les accidents majeurs provoqués par la haute technologie et la grande industrie ; ensuite comment surmonter les conflits ethniques et faire respecter les valeurs de tolérance et d'indulgence dans des sociétés de plus en plus multiculturelles ; enfin comment contrôler les manipulations biotechnologiques des micro-aspects du monde ? Tous ces problèmes témoignent du fait que l'homme n'est plus un observateur neutre du monde, mais un participant plus ou moins informé.

Cette participation a aussi une dimension économique. Vittorio Mathieu analyse l'économie comme science du rôle de l'argent à deux niveaux : celui de l'individu et celui de la société. Dans le commerce entre des individus chacun peut poursuivre ses propres intérêts, mais pour la société le résultat du commerce est autre chose que les fins poursuivies par chaque agent économique. Le résultat est une distribution des biens et des services qui n'est pas visée par les achats particuliers et les ventes particulières des commerçants. Cependant on peut effectuer des interventions non-économiques, c'est-à-dire politiques dans la vie économique. Mais les effets de ces interventions sont limités, car il n'y pas d'économie sans la liberté de chacun de disposer de son argent comme ressource à transformer en travail en vue d'obtenir des produits ou des services. Posséder l'argent est une liberté de choix, c'est avoir un pouvoir qui permet d'attendre, donc d'avoir du temps disponible. Et si la valeur de l'argent baisse par inflation, c'est certes un fait économique, mais sa cause ne l'est pas. L'inflation est l'effet d'un brigandage qu'il faut combattre comme un mal politique. Par contre dans un État de bien-être les citoyens sont assurés d'un minimum d'argent pour pouvoir disposer librement de leur propre temps.

Or le monde économique n'est pas la seule sphère de liberté. Tomonobu Imamichi, philosophe de l'éco-éthique visant la vie bonne dans une conjonction technologique, analyse la capacité de l'expérience esthétique d'ouvrir un autre monde que ceux qui résultent de l'anéantisation de « la plénitude de l'instant » par la technique de l'accélération du temps. Par l'expérience de l'art l'homme peut en effet s'opposer à deux visions contraires du monde sans place pour la liberté humaine : celle d'un monde mécano-mathématique, et celle d'un monde d'une inconscience animale.

Selon la tradition de l'Orient cette expérience esthétique est liée à une expérience religieuse qui autorise l'extase qui permet à l'esprit humain de se distancier du monde de l'efficacité de la technique. Ainsi pour Imamichi l'art est-il l'ouverture vers « l'horizon de l'être éternel que la technique veut remplacer par sa fonction continue ».

Aussi il ne faut pas oublier que la politique fait également un monde où de grandes tâches se présentent à nous. Sur la base d'une conception du monde de la politique comme un monde de pouvoir Francisco Miró Quesada C. souligne que les grandes questions politiques ne sont pas solubles sans des visions du monde qui expriment des systèmes de valeurs. Par exemple, pour s'opposer à la guerre, au racisme, etc., et pour respecter les droits de l'homme, une vision purement pragmatique du monde ne suffit pas ; il faut une réflexion philosophique qui fait valoir un idéal rationnel de la vie.

C'est cette idéal que retrouve David Rasmussen dans la vision kantienne de l'histoire « au point de vue cosmopolitique » et dans l'idée du raisonnabilité (*reasonability*) selon *Le droit des gens* (*The Law of Peoples*) de John Rawls. Rasmussen considère que prescrire, comme l'a fait Rawls dans ce livre, une coexistence raisonnable (et pas seulement rationnelle) entre sociétés dites libérales et sociétes non-libérales implique un cosmopolitisme faible, bien que Rawls ne se voie pas comme cosmopolite. Il s'agit en effet d'une vision du monde fondée sur le jugement esthétique que les êtres humains partagent en étant des êtres sensibles et qui fait qu'on peut adopter des principes fondamentaux communs de la vie pratique et politique et en dégager l'obligation de vivre ensemble. En ce sens le *sensus communis* apparaît comme un concept cosmopolite, qui fonde une mondialisation politique du monde.

Ainsi, ce livre s'inscrit dans le débat actuel sur la mondialisation qui pour le bien et pour le mal intègre toutes les manifestations de la vie humaine dans un système global. Le danger de cette transformation globale du monde vient du fait que tout pouvoir local, régional et national risque d'être neutralisé au profit d'un pouvoir techno-économique anonyme et indéterminable. En introduction, je cherche à saisir le sens de cette mondialisation dont dépend le destin de notre planète.

Nous remercions le Centre national de la recherche scientifique (Paris), le Conseil international de la philosophie et des sciences humaines (UNESCO), et l'équipe de Kluwer Academic Publishers (Dordrecht) pour leur soutien à cette publication.

Peter KEMP.

THE GLOBALIZATION OF THE WORLD

Peter KEMP
President of F.I.S.P. (Fédération internationale des sociétés de philosophie)

When this work was prepared, the theme of « the world » could still be understood as particularly philosophical, and speaking of worldhood (in French : *mondialité*) could be considered as belonging exclusively to the vocabulary of philosophers such as Martin Heidegger and Maurice Merleau-Ponty. But today, at the beginning of the 21st century, everyone talks about globalization (*mondialisation*).

Globalization is of course considered to be economic above all. One thinks of the global economy, of the global market in which the importance of local, regional, and even national political authorities is reduced more and more, to the benefit of transnational financial powers. But this economic globalization is unthinkable without a technological globalization of the means of communication ; and thus globalization, for better or worse, also comes to encompass all of culture.

For this reason, this all-encompassing globalization requires a globalization of ethics ; that is, an affirmation of universally valid norms by which the creation and development of worldwide technology – which not only condition the functioning of the global economy but also present a challenge to political and cultural life – can be evaluated and judged.

1. THE DEVELOPMENT OF GLOBALIZATION

It was the Canadian Marshall McLuhan who in 1964 proposed the conception of the world as a global society or « global village », a result of media and especially television, which made our eyes and ears

P. Kemp (ed.), World and Worldhood, 15-28.

henceforth capable of apprehending the entire world[1]. The decisive event that inspired this idea of a worldwide society was the funeral of president John F. Kennedy, which McLuhan followed, with the entire world, on the screen of a television. The beginning of the Vietnam war, displayed live on television, also played a role for this new conception of the world. It is remarkable that this first conception of globalization issued from reflection on the importance of a technology, not from an argument about the economy. Later, television companies such as CNN and BBC fully realized McLuhan's vision of an electronic world that united humanity in one global point of view on reality.

The conception of globalization as technological creation has endured for a considerable length of time. In 1969, Zbigniew Brzezinski, then professor of the University of Columbia and the future advisor of president Jimmy Carter, spoke of the American superpower as the « first global society in history »[2] because it was created through the advent of the telematic revolution. This society, according to Brzezinski, succeeded in proposing to other nations a « global model of modernity », which asserted the same normativity over the entire planet.

It was later Theodore Lewitt who, in a 1983 article in *Harvard Business Review*, first spoke of financial globalization using the expression *the globalization of markets*, which signifies that, by selling the same things, in the same manner everywhere, all the world's markets converge. Lewitt opposes this globalization to the relation between the markets of rich countries and those of less advanced countries, according to which multinationals sold to poor countries products that had become obsolete in rich countries. In 1990 Kenichi Ohmae published *The Borderless World : Power and Strategy in the Interlinked Economy*. In this book he extends the notion of globalization to every link in the chain of creation of economic value, whether it be research development (R-D), engineering, production, marketing, service, or finance. This process results in global integration, where corporations belonging to the same group manage their R-D, finance their investments, and recruit their personnel on a global scale. Globalization thus specifies a form of management of the large multinational corporation that is fully integrated with the global order.

---

[1] Marshall McLuhan, *Understanding Media : The Extensions of Man* (Toronto : McGraw Hill, 1964).

[2] Cited by Armand Mattelart, « La nouvelle idéologie globalitaire », in *La mondialisation au-delà des mythes*, éd. Serge Cordellier (Paris : Éditions La Découverte, 1997/2000) : 85.

The foundation is thus set by which the most internationalized businesses can establish the rules of the economy and release themselves from rules previously imposed by nation-states. As Serge Cordellier states in a book he wrote on globalization, one « passes from micro- to macro-economy, from rules of good private management to the establishment of economic politics and the construction or redefinition of national institutions »[3]. The latter become more and more powerless in the face of the strategies of large corporations. It follows that, as consumers give in to the temptations of international producers who offer them standardized products at a good price and choose a mode of production beyond all democratic control, governments and national parliaments could lose all ability to influence national economic developments. Transactions now pass directly to the international level, where they also risk being disrupted by a casino-economy that is capable of bringing about the most extreme losses of capital.

Yet the development of this globalization has transformed the notion of globalization itself. The world is no longer a global village in which everyone has the same experience at the same time, in worldwide transparency, but it has become an opaque network of economic and technological manipulations. This means that globalization does not entail a stronger or more profound common life among people, but an increase in segregation and an exclusion of society's most vulnerable. The abolition of borders has become not a liberation of people but a liberation of new divisions, cutting across ancient associations, between the dominating and the dominated.

The entire food market has become particularly suspect, since products lose quality by standardization, and the risks of a more and more geneticized production are less and less controllable.

It is true that globalization also carries with it positive aspects, accelerating the growth in the production of certain works of quality, and above all increasing the exchange of information and innovations that encourage the increased circulation of ideas and the proliferation of inventions. Not everything must be rejected. But without control of this globalization, one gives priority to the functioning of the market as opposed to all other matters : social progress, improved health, cultural education, environmental protection, and sustainable and viable development.

---

[3] *Ibid.*, 15.

The consequences of globalization are never purely economic, because the global economy is conditioned today by technological development and above all by the development of the means of communication of persons and information – especially by the means of rapid transportation, e-mail, and the internet.

The economy, or the symbolic exchange of product and service values, is in fact nothing without the technologies of communication. Even pure economic speculation does not intervene in the market without being conveyed by the telematic system. And even if this system does not render the world transparent, as McLuhan believed, this theorist of the « technological extensions of man » had reason to conceive of globalization as an effect of electronic communication. The universal aspect of technologies is no longer only the fact that a tool can become universally usable, or that the combination of science and *techné* that fundamentally constitutes modern technology is the condition for the project of a worldwide technological civilization, but also the fact that communication among people is capable of integrating all members of humanity and of transforming the individual of the « dwelling » into the individual of the world.

That economic-technological globalization encourages one to understand oneself as a global being is a promising development, but one that also constitutes a great danger.

2. THE DANGERS OF GLOBALIZATION

Since globalization is conditioned by technological development and permits the individual to affirm himself as a global being who has created himself by means of technical thinking, the individual of the globalized society can be tempted to understand his own thinking as purely instrumental.

The danger of this notion of thinking has often been noted by philosophers who have shown that, if thinking is only an instrument among others, man would be incapable of understanding himself. Everything is a means, but for what ? This question would have no response, for one would not conceive of a human being in this case as other than a thing – not, for instance, as an end in itself, as Immanuel Kant did.

It is this reduction of thinking that made possible the ideology of expertise I examined in my ethics of technology, *L'irremplaçable*[4]. It is the idea that all the questions one could ask have a technical solution, if a solution exists. Thus the political questions concerning social justice, that is, the equitable distribution of goods and burdens, services and obligations, are considered problems that must be resolved by means of technical measures. Therefore, there no longer remain any choices based on conviction and point of view ; no matter what difficulty there is, it is in principle surmountable by science and calculation.

This is why specialists are called in to resolve questions and thus to intervene as experts – that is, specialists who leave their proper domain in which their expertise is sufficient in order to enter into a different domain in which conceptions of humanity, society, and nature conflict and should have the last word.

The danger of what Herbert Marcuse called the thinking of one-dimensional man[5], of which the ideology of expertise is a manifestation, increased yet again with technological globalization. For the technical individual, who understands himself as a global individual by means of his participation in the process of globalization, is not only blind with respect to himself, but also with respect to communication with others.

People today do not really know what communication is. They often believe that it is an exchange of information, as though the living individual were a computer connected to the internet and exchanging information with other machines. But the computer can never become something other than a means of processing data and transmitting information, and this computer is nothing without the human being who uses it as a means. Moreover, there is no means of communication without at least two living beings who live together and exchange information for this common life. This is why the most profound communication concerns this very community ; it addresses itself to the other in order to affirm him and say, maybe even without words : You have value for me. True communication is the affirmation of existence ; the exchange of information is secondary and sometimes not even necessary.

---

[4] Peter Kemp, *L'irremplaçable, Un Éthique de la technologie* (Paris : Cerf, 1997) : Ch. 4. Cf. Pierre-Philippe Druet, Peter Kemp, Georges Thill, *Technologies et sociétés* (Paris : Éditions Galilée, 1980) : Ch. 4.

[5] Herbert Marcuse : *One-Dimensional Man* (1964), 2nd ed. (Boston : Beacon Press, 1992).

Thus we find ourselves today in a paradoxical situation in which, thanks to globalization, humanity has never had a greater capacity for communication, but also in which, in this world become global, the comprehension of communication as affirmation of the existence of others has never been so mediocre.

If the danger entailed by globalization is the disappearance of the sense of true communication, then we also risk losing control of technologies, which thereby become dangerous for their users – for consumers and for those neighboring factories that employ high-risk technologies in their production.

Certain major technological accidents – those of Seveso, Bhopal, and Chernobyl among the most serious – have been signs of alarm, arousing the public and making it attentive to the danger of those technologies that are capable of escaping from all control and thus resulting in catastrophe.

Today, however, one speaks less of possible accidents in the chemical and nuclear industries and more of the dangers that the food production industries represent, due to their development and use of biotechnologies on plants and animals. Where this production is globalized, an accident like that of the mad cow can eventually have enormous consequences, by spreading to all the corners in which the market touched by this accident extends. The effects of biotechnological and especially genetic accidents can also spread like AIDS has, since the need for nourishment is no less strong than the sexual need.

The control of technologies is today more difficult than ever, since globalization means that all decisions important for technological development are : a) assumed by the centers of power concealed in international businesses ; b) substitute for political decisions ; and c) thereby threaten all democracies and their power of control over this development.

3. The Globalization of Ethics

These dangers of globalization constitute a challenge to ethics, or to the conception of the good way of life. For it is unquestionably in the daily experiences with close relatives and in social groups (family, work teams, community, etc.) that human beings experience good and bad, and forge visions of the good life by which they can justify moral judgments.

How do these visions apply to the social relations that technical-economic globalization requires and institutes ? Are not all these visions of local origin inadequate to the worldwide systems that invent their own rules and aim at imposing their own norms on life, be it local, regional, or national ?

The same question can be posed if one supposes that norms are often instituted in a group or society by the will of a powerful person or a few rulers whom people fear and obey. For the origin of the moral can also be a local origin – a group or society – and one might wonder how norms that are valuable in a limited context could attain any validity in a global context.

However, if we have been able to observe an increasing globalization of the economy on the basis of technologies of communication that have become worldwide, we can also follow the development of ethics as a process toward the globalization of ethics. Moreover, it is well-known that the criterion of universal validity has been applied to ethics by European philosophers for several centuries, and particularly by Kant. I do not intend, however, to trace here the history of the globalization of ethics. I prefer to attempt to show a logic of globalization according to which the original vision of the good life can serve as the basis of a development that, by means of various stages, results in an ethical vision concerning all life on earth.

First, the ethical must necessarily be narrated, because it arises from an experience of the course and passing of life, in which the will to good and the tendency toward evil have been manifested. Narrative is necessary because narration is the only form of language that progresses as life progresses, and so the only language that can grasp the sense of action, be it good or destructive. The narrative takes time as life takes time.

But should we not suppose that many moral norms are originally the expressions of the will of a powerful master ? Certainly. The fact that people obey a person of power does not indicate that they do it out of respect for ethical ideas or for the moral law of their own practical reason. They can do it out of pure fear or out of need for protection, which the powerful they obey can offer, without being motivated by the recognition of a conception of the good life. In other words, norms do not have true validity and are not truly moral if they are not justified by a conception of ethics, based on a narrated vision, which testifies to a fundamental experience of the good life.

From the moment at which a person has the courage to make use of his own reason and to refuse to follow the one who says : « Do not reason, obey ! » – it is at this moment that he has succeeded in emerging from the state of immaturity (*Unmündigkeit*), as Kant writes in his essay, *What is Enlightenment ?* It is the moment at which he no longer wants to obey a moral norm without simultaneously comprehending how this norm can be justified as aspiring toward the good life. And it is therefore the moment at which the narrative vision becomes essential as the foundation for this justification. But the narratives that claim to speak of the good life are numerous, and often even indirectly describe its destruction ; we live in an ocean of stories. Thus we can ask ourselves : What are the most important or most decisive narratives ?

Before answering this question, a clearer conception of the relation between moral norms and the narrative visions of ethics must be developed. This relationship is rarely direct, or without intermediaries.

The primary intermediary is the ethical idea one draws out of certain narratives. Indeed, certain narratives show themselves to be more meaningful and resistant to repetition than others. From them we can glean content that, although initially abstract, can be rediscovered as common through shared narratives.

The idea of care for the weak has been drawn out in our culture as the central idea of our narrative universe, its center of gravity. Expressed under the form of an ethical injunction, it is known as the commandment of love for one's fellow human beings. It is the love that the Christians celebrated as *agape*, or community through love. In medieval culture it becomes charity, *caritas*. And today it is called « care » in English, « Fürsorge » in German, and « sollicitude » or « souci pour autrui » in French.

It is true that these different names do not indicate precisely the same thing, since they express notions that are marked by the particular context of meaning to which each belongs. For example, the Latin term *caritas* has assimilated classical Greek culture's idea of friendship. We can maintain, however, that there exists a fundamental conception of solicitude that is the result of a series of ideas by which human beings, in the course of our culture's history, have assembled the content of the most expressive and persuasive narratives of good and evil.

Must we conclude from this that ethical ideas are relative ? Yes or no ?

Yes, insofar as it is true that these ideas would never have formed if they had not appeared in a determinate culture. Yes also in the sense in which an idea clearly distilled from a particular culture can indicate more than an idea whose abstract content one finds quite easily in several cultures. In this way the Judeo-Greek idea of care for others indicates more than the Golden Rule : The idea of care concerns the concrete help of others, whereas the Golden Rule is limited to stipulating that one do to the other what one wishes the other would do to oneself (or does not do to others what one does not wish others to do to oneself).

No, because nothing prevents that an idea first appearing in a particular culture could progressively acquire universal validity. This depends above all on dialogue among cultures, and on the trustworthiness that the idea is capable of representing to the eyes of those belonging to cultures in which this idea has not been central from the start. Unfortunately, Europeans have often not succeeded in practicing the love they have preached to others.

Declarations and international conventions on the rights of man, however, testify to the fact that ethical ideas that contain some concrete content can – in prescribing respect for the freedom and dignity of people – be affirmed globally and obtain universal validity. These rights become real expressions of care for others, insofar as they are not only affirmed as my or our rights, but, with Emmanuel Levinas, as « the rights of the other »[6].

This reflection has already led us to recognise that the ethical idea cannot remain too general when it is a matter of translating the narrative ethical vision into concrete norms. We need more precise notions of what must be taken into account if the good life is to be protected against violence and repression. I call these notions « ethical principles ». They are not concrete rules of action, but express valuable aspects of the good life that must be preserved.

These aspects are not necessarily regarded as the only aspects of the human person's life, but they can be aspects of our biological and social life, and even of our life with other living beings ; they can be aspects of the whole living world, insofar as it is a system of resources and of life.

---

[6] Emmanuel Levinas, « Droits de l'homme et bonne volonté », in *Entre nous, Essais sur le penser-à-l'autre* (Paris : Bernard Grasset, 1991) : 231-235.

I have already mentioned the principle of respect for freedom, that is, for autonomy, and the principle of respect for dignity. Let us add here the principles of respect for integrity and respect for vulnerability.

Let's clarify the role of these principles.

The principle of autonomy is the principle of respect for self-determination, the basis of classical declarations of the rights of man, that justifies all the norms prescribing freedom of expression, action, choice, election, etc. In medicine it is the basis of the rule of clear consent.

The principle of dignity is originally the principle of respect for a person invested with social authority. That is, dignity originally concerns others, and is thereby unlike autonomy, which originally concerns the self. This is why if one transforms this idea of dignity as a quality into a universal principle for understanding the value of the human being as such, then dignity can be attributed not only to men and women as rational beings, but also to human beings who, for example, do not yet or no longer have an autonomous will, and who for this reason are incapable of controlling their own lives.

The principle of integrity has today come to have much relevance in medical ethics, and in bioethics in general, insofar as it is essential in order that the integrity of living beings not be undermined. The Latin term « integrare » comes from *tangere* (to touch) and the privative *in.* Integrity literally means that which one must not touch, damage, or alter, but which should be respected and protected. It also refers to that which constitutes a totality, a coherence ; it is in a certain sense a coherence upon which one must not infringe. Insofar as ethical rules require respect for the dignity and integrity of the person, dignity and integrity are conceived as fundamental aspects of human life that must not be manipulated or violated. These aspects constitute the coherence of life, *die Lebenszusammenhang* – not only of the person's physical life, but principally the coherence of this person's life in space and time – a life created by the attachment to a natural and social environment, by the memory of important events in it, and by the interpretation of a lifespan.

The principle of vulnerability is a more modern concept, and tied to the concept of integrity because it expresses care regarding the fragility of an intact totality that can be destroyed. Notice that in the ethical and legal debate – and in the work of philosophers like Jürgen Habermas, Hans Jonas in his *The Imperative of Responsibility*, or Emmanuel Levinas, who often speaks about the vulnerability of oneself and « the absolute nudity » of the face of the other – the idea of vulnerability is usually not simply used in a neutral description of the human condition. On the contrary, it possesses a normative sense that calls for attentive care in regard to the biological, social, and cultural (and in Levinas, moral) vulnerability that characterizes living human beings in the modern technological world.

We should add that these four principles express care for others as irreplaceable in the context of social justice, a context, that is, of solidarity and responsibility.

It is in this way that, without losing the vision of the good life as the ultimate justification of moral norms, ethics is globalized. In the face of economic and technological globalization, ethics becomes a demand for democratic and moral control of the anonymous and complex powers of corporations and international banks, which threaten the functioning of democratic political life at all levels of power.

This ethics of the irreplaceable goes beyond all moralities of proximity and toward a morality of distance that concerns the entire world. Thus the globalization of ethics signifies a limitless extension of care for the other, in the space where all living beings of the globe are concerned, and in the time when the viable and sustainable development of future generations is envisaged.

4. THE CITIZEN OF THE WORLD

Today's individual has become a global individual, and the ethics of this individual has been globalized. At the social level this suggests, I believe, that this individual must become a citizen of the world.

The conception of the citizen of the world arises in Stoicism and is revived by Kant, who, on this point as on many others, was inspired by the Stoics. Yet the Stoic idea cannot be resumed in our time without further development. Even the Kantian conception of *Weltbürger* must be adapted to new conditions. The first in European culture who named himself a « citizen of the world » was Diogenes the Cynic, a Greek philosopher who was born in Sinope and lived from 413 to 327 B.C. (that is, contemporary to Plato and Aristotle). If anyone asked him where he came from, he would respond : « I am a citizen of the world ». In so doing, he refused to let himself be determined by his local origin. He wanted to determine himself and thus to affirm his autonomy. But who was he in his own eyes ? We do not know, for he left no written trace ; we know him only through the accounts of others.

Three hundred years and a century later, respectively, Cicero and Seneca developed the idea of *societas generis humani*, the society of the human species, as the most vast conception of the society to which they belonged. Each man thus belonged to two societies : the one in which he was born and the global society. In this way even the head of state could be understood as a member of a kingdom greater than his own state. It was true of the Roman emperor Marcus Aurelius (121-180), who was also a Stoic philosopher and wrote in his *Meditations* : « I am a member of the system of rational beings »[7].

This cosmopolitanism was, however, of a purely spiritual character. It indicated a community of thought, and for the Stoics the inner life was the true human life. « Look within, » as Aurelius said. « Within is the fountain of good, and it will ever bubble up, if thou wilt ever dig »[8].

With Kant this idea of the citizen of the world became more concrete, but it nevertheless remained an insubstantial idea. The necessity of this idea appears clear only today, when we better understand its pertinence. The Kantian idea of the citizen of the world is quite superficial, although in his *Education* he contends that it must become the ideal in the education of children : « The basis of a scheme of education must be cosmopolitan »[9].

Today, however, this idea is absolutely essential ; it is the predominant idea behind thought and action, and indispensable if our earth is to survive as a dwelling for human beings in the age of globalization.

In the « Doctrine of Right » of *The Metaphysics of Morals*, and in other writings (especially *Perpetual Peace*), Kant makes a distinction between the law of peoples (*das Völkerrecht*) and the cosmopolitan law (*das Weltbürgerrecht*)[10]. The law of peoples concerns the rules of war, within certain limits. It presupposes that war is inevitable. The cosmopolitan law, on the other hand, assumes the moral law that each

---

[7] Marcus Aurelius, *The Meditations of Marcus Aurelius*, tr. George Long (Mount Vernon : Peter Pauper Press, 1957) : Book 7, §13.
[8] *Ibid.*, Book 7, §59.
[9] Immanuel Kant, *Education* (1803), tr. Annette Churton (Ann Arbor : University of Michigan Press, 1960) : 15.
[10] Immanuel Kant, *The Metaphysics of Morals* (1797), §53-62, tr. Mary Gregor (New York : Cambridge University Press, 1996). See also the commentary of Monique Castillo in Immanuel Kant, *Histoire et politique* (Paris : Vrin, 1999) : 47-57.

rational being must recognise, which implies respect for all other human beings and peoples and thus prescribes perpetual peace.

The inadequacy of Kant's conception here is, in my opinion, that he cannot conceive of a relationship among states that is neither an alliance of states, which remain completely independent, nor a superstate in which national states are dissolved. He was opposed to the idea of a global government, since he feared that it would inevitably lead to a situation in which certain rulers or a single person would seize power over all others. For Kant, therefore, no international institution other than a permanent congress of states is possible, whereby the sovereign states would meet regularly, with the sole objective being to agree to maintain peace. In other words, he could not imagine a federation of states, even though he called his conception of association « a federation of free states »[11].

But a federation is a rather abstract affair if, as in Kant, it is only the people's common acceptance of moral rules that obliges them to abstain from using war to resolve their divisions. Kant's federalism is a moral idea with a minimum of political concretization. But a truly realized federation must be a collaboration between equal states (therefore the necessity of two legislative chambers and two types of government : national governments and the federal government), and it must ultimately be a collaboration of different cultures on questions that concern all of humanity. That is, it must not only consist in a renunciation of war, but also in common regulations for the environment, especially through control of technologies and of implicated economies ; common regulations for the struggle against criminality and terrorism ; common efforts for the institution of distributive justice, that is, for a reasonable distribution of goods and responsibilities among all people ; a common desire for durable and viable development ; a mutual respect for differences in lifestyles, religions, etc.

In a true political federation, there should be common regulation regarding all questions that necessarily concern everyone. And in order to create a global federal society, the affiliation of different federations must be progressively developed toward the gradual establishment of federal connections among them. In order to assure genuine peace, it is probably not necessary that this development result in a global federal

[11] Immanuel Kant, « Perpetual Peace », Section UU, Second Definitive Article, in *Kant : On History*, ed. Lewis White Beck (Indianapolis : Bobbs-Merrill, 1963) : 98.

government, and this government is perhaps not even desirable (for it risks becoming the basis of the global state that Kant, with reason, feared). But certain competencies must gradually be transferred to international authorities, while local, regional, and national differences must be simultaneously conserved and protected. A federation must indeed protect the treasures and traditions of particular states, as much as it must be an affiliation of those states.

The citizen of the world in our time presupposes a dialogue among cultures, and between cultures and civilization. By civilization I understand here a social order that has communicative, technical, and scientific structures, whereas by culture is understood a certain form of intellectual, literary, and artistic life that characterizes historical, social, and individual self-understanding.

There is today a single civilization that overshadows all others and influences all cultures ; it is the civilization of technical-scientific and economic-capitalist globalization, with all its standardizations, and which, in the last analysis, is a menace to all cultural differences.

Cultures develop as attitudes and convictions through narratives, forms of thought, and poetry and other poetic-aesthetic and poetic-ethical expressions, and this development requires unceasing renewal (new poets, artists, thinkers, etc.) in order to subsist. It is the dialogue among these cultures that is truly difficult, but essential in order that we be made citizens of world and, as Kant says, become hospitable toward strangers.

Today economic-technological globalization constitutes a challenge to ethics that can only be relieved by a globalization of ethics itself ; that is, by a sufficiently concrete conception of a citizen of the world who is aware of himself as responsible for international democratic control of economic-technological systems, and who seeks to found this global democracy on dialogue that is geared toward the mutual understanding of different cultures, and on respect for the Other – both as other and at the same time as the nature that determines and nourishes us.

Translated by Shannon Hoff.

L'HOMME DU MONDE

Rémi BRAGUE
Université Paris I, Ludwig-Maximilians Universität, Munich

Placer l'idée de « monde » parmi les problèmes philosophiques actuels n'est pas sans quelque paradoxe. On peut se demander en effet s'il ne serait pas suprêmement inactuel. Par ailleurs, le traiter dans une optique anthropologique peut sembler surprenant.

Certes, on perçoit bien une tendance chez les philosophes, ou simplement une demande à eux adressée, de ne pas se désintéresser des problèmes du monde. On entend par là un refus d'utiliser un vocabulaire trop abscons, ou même de s'enfermer dans certains problèmes purement techniques et sans pertinence pour la société. Mais, d'une part, le monde est dans ce cas pour la philosophie plutôt un public, au plus un contexte, qu'un objet ; et d'autre part, le mot est employé alors en un sens impropre, à la fois très large et très étroit. Il est très large parce qu'il déborde le domaine du cosmologique pour englober les faits humains et sociaux ; mais très étroit aussi parce que ceux-ci ne constituent qu'une partie de l'ensemble de ce qui existe, et, même si on prend en compte le globe terrestre dans son ensemble, considéré comme une unité solidaire, comme la demeure commune des espèces vivantes dont une, l'homme, a la responsabilité, qu'est-ce que la terre quand on la compare à la totalité de l'univers ?

1. COSMOGRAPHIE ET COSMOGONIE

Il me faut donc introduire ici un peu de clarté, en précisant que je prendrai bien le monde au sens où il désigne l'objet d'une cosmologie. Et pourtant, c'est cette précision même qui réintroduira le souci de l'homme au cœur de la question du monde. J'utiliserai ici le concept de

P. Kemp (ed.), World and Worldhood, 29-89.

« cosmologie » en un sens qu'il faudra préciser. Pour ce faire, je commencerai par le distinguer de deux autres, avec lesquels on le confond souvent. Je m'appuierai sur des mots déjà présents dans le langage, mais sans m'astreindre à en respecter les acceptions reçues. Je distingue donc cosmographie, cosmogonie et cosmologie.

Par **cosmographie**, j'entends le dessin ou la description (*graphein*) du monde tel qu'il se présente actuellement, en sa structure, sa division éventuelle en niveaux, régions, etc. Cette description peut, voire doit, rendre compte des rapports statiques ou dynamiques entre les différents éléments dont se compose le monde : distances, proportions, etc., mais aussi influences, réactions, etc. Elle implique la tentative pour dégager des lois régissant ces rapports. Il s'agit ainsi d'une géographie généralisée, qui, au mépris de l'étymologie, ne porterait pas que sur la terre, mais sur l'ensemble de l'univers visible.

Par **cosmogonie**, j'entends le récit de l'apparition des choses, ou, si l'on veut, le récit de la cosmogenèse. Il explique comment les choses en sont venues (*gignesthai*) à former le monde tel que nous le connaissons, dans la structure qui est la sienne aujourd'hui. La façon dont une culture donnée conçoit le monde entraîne évidemment pour elle une certaine façon de s'en représenter la venue à l'être. Une cosmogonie sert à expliquer le monde tel qu'il est imaginé ou conçu à un moment donné par un groupe donné. Par suite, on possède des cosmogonies de styles très variés, aussi variés que celui des cosmographies. Une cosmogonie peut être mythique. C'est le cas dans les récits d'émergence que connaissent la plupart des cultures dites « primitives ». Cela n'empêche pas ces mythes d'être lourds de pensée, voire d'une réflexion d'allure pré-philosophique, comme dans la *Théogonie* d'Hésiode. Les récits sur la genèse peuvent aussi consister en la reprise partiellement critique de mythes plus anciens, corrigés, voire rendus méconnaissables par leur absorption dans un autre récit, au service d'une autre doctrine. C'est le cas du récit de la création au début du livre de la *Genèse*. Un mythe peut enfin être sciemment forgé, afin d'illustrer une théorie philosophique antérieure, comme dans le *Timée* de Platon. Une cosmogonie peut être également scientifique. Elle cherche alors à se représenter à la suite de quels événements le monde, tel que nous le connaissons actuellement, aurait pu venir à se former. C'est le cas dans le *Traité du monde* de Descartes (1633), puis dans les cosmogonies d'après Newton, comme par exemple la théorie du ciel de Kant (1775). C'est enfin le cas dans l'astrophysique actuelle et ce, quelle que soit la part d'hypothèse qu'elle comporte nécessairement.

Il convient de remarquer que les contenus de ces deux concepts se sont rapprochés avec le temps, au point de presque coïncider. Les théories contemporaines conçoivent en effet le monde comme en évolution. Avec cette dimension supplémentaire du temps, décrire le monde et en raconter la formation – l'histoire et la géographie de l'univers, si l'on veut – ne s'opposent plus comme s'opposaient autrefois, d'une part, la description d'un état fixe et constatable, d'autre part, la reconstitution purement hypothétique de sa genèse, laquelle n'avait qu'une valeur heuristique. La fabrication du monde par un divin artisan (démiurge) dans le *Timée* était, selon toute vraisemblance, et comme l'avait compris dès l'antiquité la majorité des interprètes de Platon, une façon d'expliciter en les étalant dans la succession d'opérations de fabrication, des états de choses éternels. En revanche, les sciences modernes prétendent bien raconter une histoire réellement advenue : la paléontologie le fait pour le vivant, la géologie pour les matériaux qui forment notre globe, l'astrophysique, enfin, pour l'ensemble de l'univers.

2. COSMOLOGIE

À la différence du sens dans lequel je prenais les deux concepts qui précèdent, ce que j'entends ici par **cosmologie** est un peu différent de ce que ce mot véhicule à l'accoutumée. On entend par là en effet ce mixte de cosmographie et de cosmogonie dont j'ai dit que les théories récentes le rendaient nécessaire. Je préfère quant à moi réserver le terme de cosmologie à un usage particulier. J'entends par là, comme l'implique d'ailleurs le mot de *logos*, non un simple discours, mais une façon de rendre raison du monde dans laquelle doit s'exprimer une réflexion sur la nature du monde comme monde. Est cosmologique un discours, exprimé ou non (dans ce dernier cas, on pourrait parler d'une « expérience »), dans lequel ce qui fait que le monde est monde – ce que l'on pourrait en appeler la « mondanité » – n'est pas présupposée, mais au contraire, que ce soit implicitement ou explicitement, problématisée. Il faut donc que le monde soit explicitement posé, et, déjà, nommé. La présence d'un mot n'implique pas la présence d'un concept, mais son absence indique à tout le moins que ce concept n'a pas été thématisé. Nous n'entrerons donc pas dans l'examen des « visions du monde » dites primitives, dans lesquelles le mot n'est pas présent. Ou nous ne les envisagerons que pour montrer en se détachant de quoi la pensée humaine a pu formuler l'idée de monde, et lui donner son nom.

Ainsi, un élément réflexif est nécessairement présent dans toute cosmologie, alors que son absence n'a rien de gênant dans une cosmographie ou une cosmogonie, où elle serait même déplacée. Une cosmologie doit rendre compte de sa possibilité, et, déjà, de la première condition de son existence, à savoir la présence dans le monde d'un sujet capable d'en faire l'expérience comme tel – l'homme. Une cosmologie ne peut donc pas ne pas impliquer quelque chose comme une anthropologie. Celle-ci n'est pas seulement l'ensemble des considérations que l'on peut faire sur certaines dimensions de l'existence humaine – dimension sociale, économique, anatomique, etc. ; elle ne se limite pas non plus à la théorie qui cherche à dégager l'essence de l'homme ; elle englobe aussi une réflexion sur la façon dont l'homme peut réaliser en plénitude ce qu'il est – une éthique, donc.

Au sens où je l'emploie, le mot de « cosmologie » implique donc une ouverture sur une anthropologie. S'agit-il là d'une violence faite à la langue ? Pire, d'une tentative pour réduire la cosmologie à une simple « vision du monde », nécessairement subjective – que le porteur en soit un individu ou une société ?

Remarquons d'abord que le sens du mot « monde » est souvent anthropologique. On peut constater qu'il sert de plus en plus à désigner des réalités humaines : la terre dans son ensemble, considérée comme le séjour de l'humanité, l'*oikouménè*, ou la société, voire une partie de celle-ci, la couche sociale la plus distinguée (un « homme du monde ») ; enfin, le mot peut désigner un milieu (« le petit monde de Don Camillo »). Ce sens, qu'il soit ou non impropre, est pourtant devenu le plus courant. Car, dans le fond, nous n'utilisons plus guère le mot de « monde » pour désigner une réalité physique en tant que telle. Dans cet usage, d'autres mots, comme « univers » ou « cosmos », le remplacent plus ou moins. En effet, il nous arrive bien de désigner des réalités cosmologiques comme « le monde », ou comme des « mondes », au pluriel – précisément quand nous évoquons la « pluralité des mondes ». Mais nous ne le faisons que si ces réalités cosmologiques sont considérées comme habitables par l'humanité, ou imaginées comme peuplées d'êtres hypothétiques qui seraient analogues aux hommes (la « guerre des mondes »). Faut-il donc en conclure que l'usage métaphorique du terme a tellement contaminé le sens propre du mot que celui-ci est devenu incapable de servir ?

En fait, il importe de voir que le mot « monde » n'est pas lui-même libre de toute métaphore. Il n'existe peut-être pas de mot qui puisse désigner l'ensemble de la réalité physique et qui n'implique une optique déterminée sur celle-ci. On pourrait le montrer pour « univers », qui met l'accent sur la façon dont cet ensemble est tourné vers une unité, comme focalisé sur elle. Je ne m'attacherai ici qu'au mot « monde ». On en voit apparaître le caractère métaphorique si l'on se replace aux origines, ce pour quoi il nous faut faire ici un détour par l'histoire.

3. Un mot tardif

Le mot « monde », ou plutôt la série de termes que l'on peut traduire ainsi, y apparaît à une date relativement tardive. Si l'on fait finir la préhistoire et commencer l'histoire proprement dite, comme il est traditionnel de le faire, avec l'invention de l'écriture, donc vers 3000 avant notre ère, on peut dire que ce n'est qu'à la moitié de l'histoire qu'est apparu un mot susceptible de désigner l'ensemble de la réalité d'une façon unifiée. L'humanité a pu se passer de l'idée de « monde » pendant la moitié de son histoire – sans parler de l'immensité préhistorique. Les grandes civilisations fluviales ayant inventé ou développé l'écriture, en Mésopotamie, en Égypte, en Inde, en Chine, possèdent certes des noms pour la terre, non comme planète, bien entendu, mais comme domaine habité, comme *oikouménè*, séjour commun des hommes et des animaux, sens auquel elle fait face à l'inaccessible séjour des dieux, le ciel. Mais les civilisations anciennes ne semblent pas avoir eu de mot susceptible de désigner le monde en son entier, dans l'unité de ses deux composants. Ainsi, le sanskrit *loká* signifie bien « monde », mais de telle façon que c'est l'expression *lokadvaya*, « les deux mondes » (ciel et terre) que nous devrions en rigueur traduire par « monde ». Il en est de même du japonais *sekai*. Le mot de l'hébreu médiéval et moderne pour « monde » (*'ôlām*) – racine sémitique dont est issu également le terme arabe – est bien présent dans la Bible ; seulement, il n'y a justement pas encore le sens de « monde », mais celui de « temps indéfini, éternité ».

Pour que l'on puisse parler de « monde », il faut en effet deux conditions : d'une part, que l'on envisage la totalité de ce qui est, sans que rien en soit exclu, et d'autre part, que l'on considère cette totalité comme unifiée, ordonnée, bien rangée, etc. Or, les plus anciennes civilisations n'ont guère formulé le concept de totalité des choses. Là où il s'agit de penser l'ensemble, on en reste, la plupart du temps, à une énumération. Lorsque la Bible parle de la création, et qu'elle nomme le résultat de l'opération créatrice, elle nomme, comme chacun sait, « le ciel et la terre »[1]. La formule est très ancienne, au point qu'elle est peut-être la première de toutes celles qui ont désigné le monde. Nous la verrons aussi chez les Grecs.

Par ailleurs, le concept de cosmos comme ordre universel n'affleure guère à haute époque, dans les pensées antérieures à ce que Jaspers appelait l'« âge axial » (*Achsenzeit*), ce pour quoi elles ont conçu l'humanité de l'homme à partir d'autres références. Serait-ce parce que ce concept était le milieu dans lequel baignaient ces civilisations archaïques, lequel milieu ne peut pas ne pas rester invisible ? On a cru pendant un certain temps que les Égyptiens avaient trouvé cette formulation avec leur concept de maât, personnifié comme la déesse Maât. Celui-ci exprimerait la façon dont l'univers forme un tout harmonieux dans lequel l'homme doit trouver sa juste place – juste impliquant aussi bien l'idée de justesse, d'harmonie, que de justice. Mais les travaux récents des égyptologues semblent avoir nuancé fortement cette interprétation. Ils montrent en effet que, loin que l'activité humaine doive simplement s'insérer dans l'ordre statique des choses, c'est au rebours la pratique juste de l'homme qui contribue à garder le monde en mouvement[2]. Quoi qu'il en soit, la *ma'ât*, qu'elle désigne une sagesse cosmique ou non, n'est pas un nom du « monde », mais, éventuellement, de l'ordre qui le gouverne.

S'il en est ainsi, ce qui empêche le surgissement de l'idée de monde n'est pas le caractère incomplet de l'énumération des éléments de celui-ci. C'est plutôt, au contraire, que le concept (plus ou moins explicite) que se font de l'ordre les civilisations archaïques est d'une extension trop grande, en ce qu'il englobe l'action des dieux ou des hommes. La division même du tout de la réalité en « ciel » et « terre » s'effectue selon un critère implicite qui relève de l'humain : terre et ciel s'opposent comme ce sur quoi l'homme a, en principe du moins, prise, et ce qui lui échappe radicalement. Le « monde » ne peut apparaître comme tel qu'à partir du moment où ce critère est mis entre parenthèses, où la réalité est donc campée en face d'un sujet, comme indépendante de lui. Il faut, pour que le monde apparaisse, que soit rompue l'unité organique

---

[1] Cf. mon « L'expérience biblique... », 107 s.
[2] Cf. J. Assmann, *Ma'at.* (...), p. 34.

qui liait à celui-ci l'un de ses habitants, l'homme. Réciproquement, c'est à partir du moment où le monde apparaît dans son autonomie que la présence de l'homme en lui va pouvoir surgir comme problème.

4. NAISSANCE GRECQUE DU COSMOS

Il nous faut donc nous tourner vers les Grecs. Ceux-ci, encore une fois, n'ont pas connu depuis toujours l'idée de « monde », mais n'y sont parvenus que de façon progressive[3]. On peut distinguer plusieurs étapes.

Lorsque Homère (vers le Xe siècle avant notre ère) veut parler de toutes choses, il se contente d'une énumération : ciel, terre, mer, etc. C'est le cas lorsque les principaux dieux se partagent l'ensemble de ce qui est. On trouve aussi chez lui l'expression « le ciel et la terre », formule qui survivra pendant de longs siècles encore.

Hésiode (environ 730-700), procède de la même façon, mais il lui arrive d'indiquer que son énumération n'omet rien en ajoutant l'adjectif πάντα[4].

À partir d'Héraclite (environ 500), on trouve à plusieurs reprises l'adjectif substantivé « toutes choses » (πάντα), mais cette fois sans l'énumération de son contenu. Chez Héraclite lui-même, c'est le cas dans sept fragments[5].

Avec Empédocle (environ 485-425), l'adjectif passe au singulier et l'on peut nommer : « le Tout (τὸ πᾶν) ». Il en est ainsi dans deux fragments[6]. – C'est à cette étape que s'arrête la Bible hébraïque, à peu près, d'ailleurs, à la même époque, avec le prophète Jérémie (vers 645-580) et la seconde partie du livre d'Isaïe (vers 500), qui nomme « le tout » (*hak-kōl*)[7].

C'est à une cinquième étape que le monde reçoit son nom propre, pour lequel les Grecs choisirent le mot κόσμος. Son étymologie n'est pas claire, mais son sens l'est, dès Homère : « ordre », toujours dans l'expression figée κατὰ κόσμον (« en bon ordre »), ou « parure » comme la bossette du mors d'un cheval et les bijoux d'Héra[8]. Le terme désigne l'ordre et la beauté, plus précisément encore la beauté résultant de

[3] Je reprends ici mon *Aristote*..., p. 28-31.
[4] *Théogonie*, 738.
[5] DK 22 B 1 ; 7 ; 53 ; 64 ; 66 ; 80 ; 90.
[6] DK 31 B 13 et 14.
[7] *Jérémie* 10, 16 et *Isaïe* 44, 24, et cf. mon « L'expérience... », p. 108-110.
[8] Respectivement *Iliade* 4, 145 et 14, 187.

l'ordre, celle que vise encore aujourd'hui une activité qui en tire son nom, la cosmétique. Son usage au sens de « monde » semble être l'invention de Pythagore. Mais nous n'avons que des témoignages, et qui plus est, tardifs : « Pythagore fut le premier à appeler " κόσμος " l'englobement de toutes choses (τὴν τῶν ὅλων περιοχήν), à cause de l'ordre (τάξις) qui règne en lui »[9]. Parallèlement, le latin *mundus*, d'où vient notre « monde », est sans doute le même mot que *mundus*, « toilette, parure de la femme », le sens cosmologique en ayant été tiré par imitation du grec. C'est en tout cas ce qu'affirme Pline l'Ancien : « ce que les Grecs appellent *kosmos*, nous l'appelons *mundus* à cause de son *elegantia* parfaite et sans défaut »[10]. Le mot *elegantia* n'est pas qu'une préciosité, mais une référence directe à l'usage cosmétique du terme *mundus*.

Cet usage n'allait pas de soi. On rencontre à date bien postérieure, chez Xénophon, un texte où une certaine gêne se fait sentir :

> « Jamais personne n'a vu Socrate faire ou ne l'a entendu dire quelque chose d'impie (ἀσεβές) ou d'incorrect (ἀνόσιον). En effet, il ne discutait pas de la nature de toutes choses (περὶ τῆς τῶν πάντων φύσεως), comme le font la plupart des autres, en examinant comment ce que les spécialistes appellent " ordre " (ὁ καλούμενος ὑπὸ τῶν σοφιστῶν κόσμος) se tient et par quelles nécessités chacune des choses célestes (τὰ οὐράνια) advient, et même il montrait que ceux qui se soucient des choses de ce genre sont fous »[11].

Le mot est mis au crédit des σοφισταί, terme qui ne désigne pas encore les « Sophistes », avec une connotation péjorative, comme depuis Platon, mais simplement les gens compétents, les habiles, les savants. Le mot est encore perçu comme un terme technique, voire un peu prétentieux, donc étranger au bon ton. Un *gentleman* comme Xénophon se doit de prendre quelque distance envers les spécialistes.

On notera l'identification entre « la nature de toutes choses » et le *kosmos*. Le mot réapparaît plus loin dans l'entretien avec Euthydème,

---

[9] Aetius, *Placita*, II, 1, 1, p. 327 Diels et cf. Diogène Laërce, VIII, 48.
[10] Pline l'Ancien, *Histoire naturelle*, II, iv, 3.
[11] Xénophon, *Mémorables*, I, i, 11.

dans lequel Socrate développe l'argument téléologique, ainsi que dans un parallèle :

> « Les <dieux> nous donnent de bonnes choses, mais ne donnent aucune de celles-ci en se manifestant ; en particulier, celui qui met en ordre et maintient la totalité du monde (ὁ ὅλος κόσμος), dans lequel toutes choses sont belles et bonnes, et qui toujours fournit à ceux qui en ont besoin des choses inusables, saines et à l'abri de la vieillesse, et qui obéissent sans faute plus vite que la pensée, ce dieu est vu quand il fait les plus grandes choses, mais quand il ménage ces choses, il nous est invisible [...] les dieux qui sont toujours, qui surveillent toutes choses et peuvent tout, eux qui maintiennent cet ordre de toutes choses (ἡ τῶν ὅλων τάξις) inusable, à l'abri de la vieillesse, sans faute et dont on ne peut raconter la beauté et la grandeur »[12].

Dans notre texte, les choses divines, distinguées des choses humaines, sont avant tout les choses célestes. La distinction n'est pas faite entre les corps célestes – qui ne sont pas nommés dans le passage – et les phénomènes météorologiques, dont il sera question plus loin : « vents, eaux, saisons ». L'accent est mis sur l'inaccessibilité, sur l'incapacité que nous sommes d'exercer une influence sur ce que nous ne pouvons que voir – contempler.

Ce n'est cependant qu'avec Héraclite, peut-être, que l'on peut trouver formulée la décision conceptuelle qui permet l'émergence de l'idée de monde. Nous lisons, dans un très célèbre fragment, dont nous risquerions une traduction en première approximation :

> « κόσμον τόνδε τὸν αὐτὸν ἁπάντων, οὔτε τις θεῶν οὔτε ἀνθρώπων ἐποίησεν, ἀλλ' ἦν ἀεὶ καὶ ἔστιν καὶ ἔσται : πῦρ ἀείζωον, ἁπτόμενον μέτρα καὶ ἀποσβεννύμενον μέτρα » (« ce monde, le même pour tous, ce n'est ni un des dieux ni un des hommes qui l'a fait, mais toujours, il

[12] *Mémorables*, IV, iii, 13 ; *Cyropédie*, VIII, vii, 22.

> était, il est, il sera : feu toujours vivant, s'allumant en mesure et s'éteignant en mesure »)[13].

Il n'est pas aisé de voir quel est le sens exact du mot κόσμος. S'agit-il du « monde » ou de l'« ordre » des choses ? Son référent n'est pas plus clair : ce qui s'allume et s'éteint en mesure désignerait-il la parure de la voûte céleste, les étoiles, avec leur ordre et leur beauté ? ou le « feu », souvent interprété comme feu cosmique depuis les Stoïciens ? Quant à l'idée selon laquelle le κόσμος n'a été fait par aucun dieu ni par aucun homme, la deuxième négation y est évidemment la plus problématique, car qui a jamais prétendu que le monde était œuvre humaine ? On peut peut-être la comprendre sur le fond de la conception égyptienne, évoquée plus haut, selon laquelle l'ordre du monde était maintenu par l'intervention des hommes. En tout cas, on voit affirmer l'idée d'une totalité ordonnée qui se suffit à soi-même et ne requiert pas l'intervention d'une instance extérieure.

Un dialogue, sans doute tardif dans la carrière littéraire de Platon, le *Timée*, fait fonctionner le concept de *kosmos* en fournissant la première description de la réalité comme formant un tout ordonné, à la fois bon et beau. Le dialogue aura une influence que l'on ne peut surestimer. Les derniers mots en résument le propos :

> « il est né notre monde (ὅδε ὁ κόσμος), vivant visible comprenant les vivants visibles, dieu sensible, image d'un dieu intelligible, très grand, très bon, très beau, et très parfait, ciel unique (εἷς οὐρανὸς ὅδε) qui est seul de sa race »[14].

La formule pompeuse fait écho à des affirmations faites plus avant : celui qui a fait n'a pas fait deux mondes ou des mondes infinis, mais « seul cet unique ciel est né et le sera encore ». Le monde est identifié avec le ciel. Le monde, qui coïncidait d'abord avec « le ciel et la terre », est pour ainsi dire absorbé par le ciel.

On rencontre le même phénomène chez Aristote. Celui-ci distingue trois sens du mot « ciel » (οὐρανὸς) :

> « (a) la substance du dernier orbe de l'univers ou le corps naturel qui se trouve dans cet orbe ; (b) le corps qui est continu avec le dernier

[13] DK 22 B 30.

[14] Platon, *Timée*, 92 c 7-d 3 ; trad. L. Brisson, p. 229. Cf. aussi 31 b 3 s.

> orbe de l'univers ; (c) le corps qu'enveloppe le dernier orbe »[15].

On aura reconnu respectivement la sphère des fixes, les corps célestes, enfin l'ensemble de ce que contient la sphère des fixes, y compris la terre. La mutation sémantique qui permet ce dernier sens est rendue possible, voire nécessaire, par une mutation dans la représentation de l'univers. Elle n'était pas possible tant que l'on imaginait le ciel comme un plan parallèle au-dessus de la terre. Mais une fois le ciel considéré comme arrondi et enveloppant la terre de tous côtés, on peut, par synecdoque, appeler le contenu du nom du contenant : comme on appelle « une bouteille de vin » non seulement la paroi de verre mais le vin qu'elle contient, de même on peut appeler « ciel » tout ce que le ciel contient. Cela n'est pourtant pas sans conséquence, en ce qu'il ne va pas de soi que l'on puisse transférer toutes les qualités du contenant à celles du contenu. Le monde est supposé « beau et bon ». Ces qualités sont-elles compatibles ?

5. ORDRE DU MONDE ET ORDRE DE L'HOMME

À partir de ce concept grec de monde comme *kosmos* s'ouvre une dimension anthropologique. Cela semble paradoxal si, comme je l'ai supposé, le concept de « monde » ne devient possible qu'à partir du moment où l'homme est exclu de son contenu. En fait, l'idée grecque de monde contient un appel implicite à l'idée de sujet, qu'elle exige comme en creux. L'homme est bel et bien exclu de tout rôle actif dans la constitution du monde. Mais c'est justement parce que l'homme ne « fait » rien au monde, c'est parce que sa présence n'ajoute rien à aucune partie de celui-ci, et que son absence ne lui retire rien, qu'il peut apparaître comme le sujet auquel le monde se manifeste dans sa totalité. Ce fait est et restera acquis, même dans les développements les plus récents de l'idée de monde, comme nous le verrons.

Pour les Grecs, le lien du monde au sujet se manifeste d'abord en ce que les deux sont régis par les mêmes lois, de nature morale. Platon l'indique en un passage célèbre :

> « les sages disent [...] que ce qui fait tenir ensemble le ciel et la terre, les dieux et les hommes est la communauté (κοινωνία), l'amitié, la

---

[15] Aristote, *De caelo*, I, 9, 278 b 11-21.

> régularité (κοσμιότης), la tempérance, la justice, et c'est pour celà qu'ils appellent le tout que voici (τὸ ὅλον τοῦτο) “ monde (κόσμος) ”, non désordre (ἀκοσμία) ni intempérance »[16].

Il n'y a pas là transfert (métaphore) de ce qui vaut pour la société des hommes à l'ensemble de toutes choses. En fait, « monde » n'a jamais désigné une description de la réalité, mais toujours un jugement de valeur, fruit d'une sorte d'acte de foi, positif ou négatif. La science grecque avait conscience d'être non seulement une connaissance du *kosmos*, mais quasiment une constitution de celui-ci comme tel, comme *kosmos*. Cela est exprimé dans un passage de Simplicius, commentateur néo-platonicien d'Aristote (début du VIe de notre ère), qui prétend lui-même reproduire un enseignement du stoïcien Poseidonios (IIe siècle avant J.-C.) :

> « La tâche de la contemplation de la nature (θεωρία φυσικὴ) est d'examiner la substance du ciel et des astres, la puissance et la qualité de la génération et de la corruption, et, par Zeus !, elle est capable de mener des démonstrations au sujet de la grandeur, de la forme et de l'ordre de ces choses. Quant à l'astronomie (ἀστρολογία), elle n'entreprend de parler de rien de tel, mais elle démontre l'ordre (τάξις) des choses célestes, ayant déclaré (ἀποφήνασα) que **le ciel** (οὐρανός) **est véritablement un monde** (κόσμος) ; elle parle des formes, des grandeurs, des distances de la terre par rapport au soleil et à la lune, des éclipses, des conjonctions des astres, sur la qualité et la quantité <qui se manifeste> dans leurs révolutions »[17].

La liaison entre le cosmologique et l'anthropologique est rétablie. Mais elle est désormais renversée : ce n'est pas l'homme qui assure l'ordre du monde, c'est au contraire l'imitation de l'ordre préexistant des réalités physiques qui va aider l'homme à atteindre la plénitude de son

---

[16] *Gorgias*, 507 e 6-508 a 4.

[17] Posidonius, fgt. 18 (Météorologiques), p. 44 Edelstein-Kidd (2e ed.) = Simplicius, *In Physicam*, II, 2, p. 291 f. Diels. Pour l'interprétation de la phrase qui nous intéresse ici, cf. Commentaire, p. 131 : « on the basis of the preliminary acceptance… that it takes from philosophy ».

humanité. La sagesse sera, comme nous le verrons, une imitation du monde.

Le commentateur néoplatonicien Elias (Ve siècle) établit le parallèle entre l'ordre du monde et celui des mœurs de qui l'étudie :

> « Ainsi donc, que l'étudiant (ἀκροατής) soit dans sa vie rangé (τεταγμένος) et <forme> un μικρός κόσμος, afin que par l'intermédiaire de l'ordre qui est en lui, il connaisse l'unique principe du Tout (en effet l'ordre est dirigé vers l'Un, le désordre vers la pluralité), et que, par l'intermédiaire du μικρός κόσμος qui est en lui, il reconnaisse le κόσμος. Or, il sera un μικρός κόσμος si, comme dans l'univers (κόσμος), les choses meilleures l'emportent sur les moins bonnes »[18].

On pourrait retourner la formule, dans l'esprit des « philosophies du soupçon » ou de la critique des idéologies : poser le monde comme un *kosmos* est le fait du sage, qui y voit le miroir de sa sagesse à lui. C'est en tout cas ce que suggère une étonnante incise de Platon :

> « Le chœur unanime des sages affirme en effet que l'Intellect est pour nous roi du ciel et de la terre, et, ce faisant, c'est d'eux-mêmes qu'ils font un objet de vénération (Πάντες συμφωνοῦσιν οἱ σοφοί, ἑαυτοὺς ὄντως σεμνύνοντες, ὡς νοῦς ἐστι βασιλεὺς ἡμῖν οὐρανοῦ τε καὶ γῆς) »[19].

En clair, et brutalement : les sages se donnent un modèle du monde qui leur assure une domination, au moins symbolique. On songe à un parallèle bien postérieur, chez quelqu'un qui se présente pourtant comme un adversaire déclaré de Platon, Nietzsche :

> « Entfernen wir die höchste Güte aus dem Begriff Gottes : sie ist eines Gottes unwürdig. Entfernen wir insgleichen die höchste Weisheit : – es ist die Eitelkeit der Philosophen, die diesen Aberwitz eines Weisheits-Monstrum

---

[18] *In Categorias*, Préface, 6e point, CAG XVIII-1, p. 121, 23-28 Busse.
[19] *Philèbe*, 26 c 6-8.

> von Gott verschuldet hat : er sollte ihnen möglichst gleich sein »[20].

6. LE MONDE COMME OBJET DE CONTEMPLATION

Malgré tout, l'Antiquité et le Moyen Age dans leur écrasante majorité, semblent avoir associé sans hésitation la sagesse humaine et la sagesse du monde. Aristote avait donné sa forme classique à l'idée de contemplation (*theôria*), mais non sans laisser sa doctrine à ce propos étrangement inachevée. Nous apprenons à la fin de son principal traité d'éthique qu'elle constitue le mode de vie le plus élevé et le plus digne[21]. Mais il ne nous dit nulle part au juste ce que nous avons à contempler, quel est l'**objet** de la contemplation. Les penseurs postérieurs, en particulier dans la tradition stoïcienne, ont une réponse toute prête : l'objet par excellence de la contemplation est le monde, dans son indistinction d'avec Dieu. Le but de l'existence humaine est d'être un *cosmotheôros* (le mot n'est pas ancien, mais peut-être forgé par Kepler). Cette réponse était d'ailleurs préfigurée, dès avant Aristote, par des penseurs comme Anaxagore, à qui on attribue des propos selon lesquels le but de la vie humaine est la contemplation du ciel[22]. L'idée se retrouve dans la tradition juive[23].

Penser que le monde est un objet de contemplation est d'autant plus aisé qu'on lui prêtera des traits divins. Les Stoïciens le divinisent donc, en l'identifiant avec Jupiter. Pline l'Ancien écrit :

> « <Le monde> est sacré, éternel, immense, tout en tout, ou plutôt, il est lui-même tout, en dehors, au-dedans, comprenant toutes choses en soi, il est en même temps l'œuvre de la nature des choses et la nature même des choses (*<Mundus> sacer est, aeternus, immensus ; totus in toto ; immo vero ipse totum ; extra, intra, cuncta complexus in se ; idemque rerum naturae opus et rerum ipsa natura*) »[24].

---

[20] Fragment de l'automne 1887, 10 [90], *KSA* 12, p. 507 = *Wille zur Macht*, § 677.

[21] *Éthique à Nicomaque*, X, 7 - 8.

[22] DK 59 A 30.

[23] Cf. S. Pines & Z. Harvey, « Voir les étoiles et les constellations » (hébreu), *Mehqarey Yerushalayim be-Makhsheveth Israel*, 3 (4), 1984, 507-511.

[24] *Histoire naturelle*, II, 1.

C'est le traité pseudo-aristotélicien *Du monde*, marqué par le stoïcisme, qui, peut-être pour la première fois, assigne à la contemplation son objet propre : la connaissance du monde et de ce qu'il y a de plus excellent dans le monde est l'affaire par excellence de la philosophie[25].

Senèque développe largement ce thème :

> « La curiosité est en nous un don de la nature : ayant conscience de son génie et de sa beauté, elle nous a créés pour être les spectateurs d'un si merveilleux spectacle ; elle aurait perdu sa peine, si ces ouvrages si grands, si purs, si finement conformés, si brillants, riches de tant de beautés diverses, elle ne les offrait qu'au néant. Ce qui prouve qu'elle veut qu'on la contemple et qu'il ne lui suffit pas d'un coup d'œil, c'est le lieu où elle nous a placés : elle nous a établis dans son centre, en disposant autour de nous le panorama du monde ; et elle ne s'est pas bornée à mettre l'homme debout, mais, comme elle tenait à lui rendre la contemplation facile, pour qu'il pût suivre le mouvement des astres de leur lever à leur coucher et tourner son visage à mesure que tourne l'Univers, elle a dressé sa tête vers le ciel et l'a posée sur un cou flexible. Puis, en faisant progresser les signes du zodiaque à raison de six pendant le jour et de six pendant la nuit, elle a successivement déployé à ses yeux toutes les parties d'elle-même, afin que ce qu'elle lui en montrait lui inspirât l'envie de connaître le reste. Il y a en effet des objets que notre vue n'atteint pas et ceux que nous voyons nous apparaissent en raccourci : c'est bien notre œil cependant qui nous ouvre la voie de l'investigation et qui pose les premières assises de notre connaissance du vrai. Nous pouvons ensuite, par induction, passer du sensible à l'obscur et concevoir une réalité plus ancienne que le firmament lui-même : nous parvenons à découvrir d'où ces astres sont issus,

[25] *De mundo*, 1.

quel était l'état de l'Univers avant que ces différentes parties ne se fussent dissociées, quelle raison présida à l'organisation de ce chaos, qui assigna sa place à chaque chose, si c'est par un effet de leur propre nature que les corps pesants descendirent et que les corps légers s'élèvent ou si, abstraction faite de leur poussée ascensionnelle et de leur poids, quelque puissance supérieure leur prescrivit cette double loi, s'il est vrai (et c'est le meilleur argument en faveur de l'origine divine de l'âme humaine) que des particules détachées des astres à la manière d'étincelles se soient précipitées sur la terre et fixées dans ce milieu étranger. Notre pensée traverse les remparts du ciel et ne se contente pas de connaître ce qui tombe sous les sens : “ Je sonde, dit-elle, l'étendue qui s'ouvre au-delà du firmament : est-ce un abîme illimité, ou bien cet autre monde a-t-il lui-même ses bornes ? Quel est l'aspect de ces régions invisibles ? Tout y est-il informe et indistinct, ou bien, occupant un égal espace en tous sens, les choses sont-elles, là aussi, disposées dans un ordre harmonieux ? Ce monde tient-il à notre monde, ou bien un intervalle les sépare-t-il l'un de l'autre et ce monde-ci roule-t-il dans le vide ? Sont-ce des atomes qui constituent tout ce qui est et qui sera, ou bien l'universelle matière est-elle homogène et ne subit-elle de changements que dans sa totalité ? Les éléments sont-ils en lutte les uns avec les autres, ou bien, sans se combattre, concourent-ils par des voies opposées à une fin commune ? ” [...]. En somme, je vis selon la nature si je me donne à elle tout entier, si je consacre ma vie à l'admirer et à l'adorer. Or la nature veut que je fasse deux choses : agir et me livrer à la contemplation ; je les fais l'une et l'autre, puisqu'aussi bien la

> contemplation même ne peut se concevoir sans action »[26].

7. Le mal comme exception

La nature dont il est ici question n'est pas d'abord le monde dans lequel nous vivons, entourés des végétaux et des animaux, le monde où les vivants naissent et disparaissent. Le monde, aussi, où, loin de se céder de bonne grâce les uns aux autres, animaux et cités cherchent à se supplanter, ou du moins, rivalisent pour s'approprier des ressources rares. Pour les Anciens, ce n'était là qu'un aspect d'une totalité. Ce qui se passe autour de nous est moins règle qu'exception. Là où le monde est vraiment ce qu'il est, ce combat n'existe pas. Ce lieu, dans lequel la nature se montre sous son vrai jour, est le monde céleste. De la sorte, le bien et le mal se répartissent comme le « haut » et le « bas », comme la terre et le ciel :

> « Il n'y a rien de bon sur la terre, ni rien de mal dans le ciel (οὐδὲν ἀγαθὸν ἐπὶ τῆς γῆς, οὐδὲν κακὸν ἐν τῷ οὐρανῷ) »[27].

Dans l'univers intellectuel de l'Antiquité tardive et du Moyen Age, ou en tout cas dans le courant dominant de cette période, on peut trouver les traces d'une certaine attitude fondamentale, attestée de tous côtés. Elle constitue le fil rouge de l'attitude éthique de l'Antiquité tardive et de ce que celle-ci a légué au Moyen Age. Je m'exprime ici avec prudence : la tendance qui est restée maîtresse du champ de bataille n'était pas la seule. L'épicurisme ou la gnose ne voyaient pas les choses ainsi. Et les religions bibliques posaient la question autrement, comme on le verra. D'après cette attitude dominante, en tout cas, le mal qui règne ici sur la terre n'est au fond qu'une exception. La règle qu'il confirme se manifeste dans la régularité et l'ordre majestueux des mouvements célestes. C'est grâce à cet ordre que le monde mérite son nom grec de *kosmos*, ce qui veut dire, justement, « ordre », bel arrangement, totalité rangée et harmonieusement articulée, etc.

D'où un fait capital : on peut faire appel du mal à une instance plus élevée. Et plus élevée aussi en un sens cosmologique. Pour Aristote, et

---

[26] *Du loisir*, V, 3 - 6. 8.

[27] Hermès Trismégiste, fragment de Stobée n° XI, sentence n° 18 ; t. 3, p. 55 Nock-Festugière.

pour la cosmologie qui s'est imposée dans son sillage, le sublunaire est séparé du supralunaire. Il l'est, comme le mot le dit, par la barrière que constitue la sphère de la lune. En bas se trouve le théâtre de la génération et de la corruption des choses qui proviennent des quatre éléments et qui y font retour. En haut règne un cinquième élément, incorruptible, qui se meut en cercle, de par sa nature propre. En bas, les événements ne peuvent se répéter à l'identique. Tout au plus forment-ils un cycle, comme les eaux, les saisons, ou les espèces animales. Mais aucun individu ne subsiste éternellement. En haut, les astres sont éternels comme individus. En bas, tout va à vau-l'eau ; en haut, « tout n'est qu'ordre et beauté ».

Maintenant, l'endroit où tout se passe bien est incomparablement plus grand que le coin minuscule où nous nous languissons, au centre du monde, c'est-à-dire, pour les Anciens, au plus bas, en dessous de tout[28]. Le mal peut très bien se déchaîner, là où nous sommes. Mais il reste une exception presque invisible, si on le compare à l'écrasante majorité du Bien. C'est ce que dit déjà Aristote, qui exprime l'idée dans le langage du droit :

> « la région du sensible qui nous environne est la seule où règnent la génération et la corruption, mais elle n'est même pas, si l'on peut dire, une partie du tout, de sorte qu'il eut été plus juste d'acquitter le monde sensible en faveur du monde céleste, que de condamner le monde céleste à cause du monde sensible »[29].

La même idée se retrouve, par exemple, chez Alexandre d'Aphrodise :

> « Le non-être n'est pas mélangé à la plupart des étants, et, là où il l'est, il n'est pas non plus grand-chose. Mais il est peu de choses dans peu de ce qui est. Parmi les étants, il est là où il y a du non-éternel. C'est ce qui entoure la terre, et cet endroit est tout petit (ἐλάχιστος) comparé à l'ensemble du monde. En effet, si la terre, selon les astronomes, est comme un point par rapport à l'ensemble du ciel, et si le non-être se trouve

[28] Cf. mon « Le géocentrisme (...) ».
[29] *Métaphysique*, Γ, 5, 1010 a 25 - 32.

autour d'elle et en elle, il concerne bien peu de choses »[30].

Elle est encore chez Avicenne :

> « La totalité de la cause du mal ne se trouve que dans ce qui est en dessous de la sphère de la lune, et l'ensemble de ce qui est sous la sphère de la lune est insignifiant par comparaison avec le reste de l'être, comme tu sais »[31].

Elle est enfin puissamment orchestrée chez Maïmonide, dans sa polémique contre le médecin et libre-penseur Râzî :

> « si l'homme considérait et concevait l'univers, et s'il savait quelle petite place il y occupe, la vérité lui deviendrait claire et manifeste. En effet, cette insigne folie que proclament les hommes touchant la multitude des maux qu'il y aurait dans l'univers, ils ne la professent ni à l'égard des anges, ni à l'égard des sphères et des astres, ni à l'égard des éléments et des minéraux ou plantes qui en sont composés, ni à l'égard des différentes espèces d'animaux ; mais leurs pensées ne se portent que sur quelques individus de l'espèce humaine [...]. Mais ce qui est la vraie manière d'envisager (*i'tibâr*) la chose, c'est que tous les individus de l'espèce humaine qui existent, et à plus forte raison ceux des autres espèces d'animaux, sont une chose sans aucune valeur (*qadar*) par rapport à l'ensemble immuable de l'univers [...]. <L'homme> ne doit point se tromper et croire que l'univers n'existe que pour sa personne. Selon nous, au contraire, l'univers existe à cause de la volonté de son Créateur, et l'espèce humaine y est bien peu de chose par rapport au

---

[30] Alexandre d'Aphrodise, *De anima*, p. 171, 28 - 34.

[31] *Šifâ'*, *Métaphysique*, IX, 6, arabe, p. 417, 4 s. ed. Anawati ; tr. fr. p. 151. Cf. aussi p. 422, 9-13 ; tr. p. 155. Le passage est cité par Thomas d'Aquin, *De veritate*, q. 5, a. 4 c, p. 96 b.

> monde supérieur, je veux dire, aux sphères et aux astres »[32].

Dans cette cosmologie, l'action humaine ne se déroule que dans un coin perdu du monde. Ce n'est que dans ce minuscule trou dans l'ordre cosmique que l'on peut et doit distinguer entre le bien et le mal et, espérons-le, choisir le bien. Partout ailleurs, le choix en faveur du bien a été fait depuis toujours. Tout « devoir-être » (*Sollen, ought*) est cerné par un gigantesque « être » (*Sein, is*), qui le relativise, voire, le rend négligeable. Et cet « être » s'identifie avec le bien, *ens et bonum convertuntur*. Il se pourrait que prendre conscience de ce fait puisse nous aider à comprendre pourquoi les Anciens n'ont guère formulé la distinction entre être et devoir-être, qui affleure pour la première fois chez Hume[33], mais que toute morale fait implicitement.

8. LE MONDE COMME MODÈLE MORAL

Jusqu'à une époque déterminée de l'histoire de l'esprit humain, la nature était une des sources de la loi morale. Je n'entends pas par là les tentatives pour fonder la morale sur la « nature humaine ». Car, dans ce cas, « nature » veut dire avant tout « essence », telle qu'elle est saisie et élucidée par une ontologie de l'humain, autrement dit, par une anthropologie philosophique. La nature de l'homme est alors ce que l'homme est en son fond, ce qui constitue sa forme la plus achevée. Et il s'agit de réaliser concrètement la perfection de l'humanité de l'homme par une praxis qui rend possible le déploiement de cette essence, voire, qui l'exige.

Il en est tout autrement dans la façon de voir que j'essaie ici de reconstruire. La nature qui déterminait l'éthique était bien ce que l'on entend habituellement par « la nature », à savoir l'ensemble des choses que l'homme ne produit pas (par sa *poïèsis*), qu'il ne « fait » pas non plus (par sa *praxis*), mais qui sont là d'elles-mêmes. La nature est par suite l'objet d'un rapport purement passif, celui qui est le cas lorsqu'un objet se donne à l'observation. Mais la nature, ainsi comprise, n'a pour nous aucun rapport avec l'éthique. Ou si elle en a un, c'est à titre de domaine d'application parmi d'autres. Elle vient même après le domaine principal

---

[32] *Guide des égarés*, III, 12, arabe, p. 318, 24-29, 319, 4-6. 9-12 Joël ; p. 68-70 trad. Munk.

[33] Hume, *Treatise on Human Nature*, III, 1, 1.

de l'agir moral, qui est, bien entendu, les relations inter-humaines. Pour relativiser cette évidence, je voudrais rappeler à la mémoire quelques textes. Ils illustrent une façon de voir le monde d'après laquelle l'homme peut, voire doit emprunter le critère de son action à la nature. La nature est alors source de moralité.

Je dis bien : la nature, non pas la science physique ; la nature comme objet d'étude, non l'étude de la nature comme activité d'un sujet. La valeur éthique de la physique ne vient pas du processus humain de connaissance, mais de la nature elle-même. Il ne s'agit pas de rappeler que l'étude de la nature peut nous entraîner à la pratique de certaines vertus, parce qu'elle les éveille ou parce qu'elle les implique. Ainsi, l'effort pour percer à jour les mystères de la nature peut nous former au courage ; la réflexion sur le merveilleux ordre des choses peut nous mener à l'admiration et à la reconnaissance envers le Créateur – que celui-ci soit un Dieu personnel ou l'âme inconsciente du monde ; considérer le peu de place que nous occupons dans le monde peut nous apprendre l'humilité, etc. Tous ces éléments ont été illustrés dans l'histoire de la pensée, avec les couleurs les plus différentes : par Pascal, mais aussi par Spinoza, Goethe, etc. Et les physiciens d'aujourd'hui reprennent souvent cet hymne à la valeur éducative de la physique.

On trouve ces idées dans les textes anciens. Mais on y trouve aussi des passages dans lesquels il s'agit clairement de la nature comme objet d'étude, non de son étude comme activité. Ainsi, Cicéron :

> « la connaissance des choses célestes apporte une sorte de modestie à ceux qui voient combien grande est chez les dieux la modération, combien grand est l'ordre, <elle apporte> la grandeur d'âme à ceux qui contemplent les œuvres et les actions des dieux, <elle apporte> aussi la justice, quand on sait quelle est la décision du guide et seigneur suprême, quel est son conseil, quelle est sa volonté »[34].

Citons encore Simplicius qui affirme que la nature est un modèle de justice : l'étude de la nature contribue

> « à la justice, dans la mesure où elle montre que les éléments et les parties du Tout cèdent les uns aux autres, se contentent de leur place,

[34] *De Finibus*, IV, v, 11 s., p. 124 Schiche.

> respectent l'égalité géométrique, et, de ce fait, se tiennent à l'écart de l'avidité »[35].

Dans un texte parallèle, Averroès, qui s'appuie sans doute, lui aussi, sur le commentaire perdu d'Alexandre d'Aphrodise, souligne que le modèle moral se trouve dans les choses elles-mêmes. À propos de la vertu de justice, il écrit que les sages sont justes :

> « car ils savent que la nature de la justice réside dans l'essence des choses, c'est pourquoi ils souhaitent imiter cette nature et acquérir cette forme »[36].

La contemplation de la nature inspire la justice parce que la nature se conduit avec justice, sans que le texte nous dise en toute clarté dans quel domaine cette justice se manifeste plus spécialement. Il peut en effet s'agir des quatre éléments qui se changent les uns dans les autres, dans l'esprit du fragment d'Anaximandre[37] ; ou de la disposition concentrique des parties du *kosmos*, dont aucune n'empiète sur la voisine ; ou de l'ordre des phénomènes astronomiques, etc. Dans le même sens, Farabi voit entre les éléments un cycle parfait : comme la matière ne peut réaliser en même temps toutes les formes, chacune cède sa place à l'autre, à tour de rôle[38].

9. L'IMITATION DU MONDE

Maintenant, cette vision du monde n'est pas seulement une description statique de la réalité en sa structure hiérarchisée. Elle contient aussi un modèle de la pratique morale. L'appel au bel ordre des phénomènes célestes n'est pas uniquement théorique. Il est aussi une figure de ce qui est demandé à l'homme, le plus noble des êtres qui sont relégués dans la contingence du sublunaire. L'ordre du monde supérieur est pour lui le modèle de la conduite correcte de la vie. C'est par l'imitation de la parfaite régularité des allures célestes que l'homme pourra réussir à

---

[35] *In Physicam*, (*CAG*, IX), p. 4, 23-26 Diels. Le texte contient une allusion transparente à Platon, *Gorgias*, 508 a 6.

[36] *Commentaire moyen à la Physique*, Préface, p. 2 C édition des Juntes. Cf. S. Harvey, « The Hebrew Translation of Averroes' Prooemium to his Long Commentary on Aristotle's *Physics* », *PAAJR* 52 (1985), 55-84, cit. p. 66, 22 s.

[37] DK 12 B 1.

[38] *as-Siyāsa al-madaniyya*, p. 59, 13-60, 2 Najjâr. Cf. aussi *Cité idéale*, ch. 19, p. 72/55 s. ; 74 s./58.

mettre de l'ordre dans sa propre vie sublunaire. Il pourra ainsi, pour transposer dans le registre pratique la célèbre formule de la « révolution socratique », « faire descendre le ciel dans les cités ». L'idée est exprimée, sans doute pour la première fois, par Platon :

> « le dieu nous a découvert et donné la vue, afin que, ayant observé dans le ciel les révolutions de l'intellect, nous les utilisions, en les rapportant aux révolutions en nous de l'intellect ; ces révolutions sont apparentées, même si les nôtres sont troublées alors que les autres sont exemptes de trouble. Ce n'est qu'après avoir étudié à fond les mouvements célestes, après avoir acquis le pouvoir de les calculer correctement en conformité avec ce qui se passe dans la nature et après avoir imité les mouvements du dieu, mouvements qui n'errent absolument pas, que nous pourrons stabiliser les mouvements qui en nous ne cessent de vagabonder »[39].

Cicéron reprend l'idée :

> « Je crois que les dieux immortels ont semé les âmes (*animos*) dans les corps humains, afin qu'il y eût quelqu'un pour s'occuper des choses de la terre et qui, en contemplant l'ordre des choses célestes, l'imitent par la mesure et la constance de leur vie (*caelestium ordinem contemplantes imitarentur eum vitae modo atque constantia*) »[40].

L'idée qu'exprime le texte du *Timée* a connu une postérité immense dans l'Antiquité et le Moyen Age, que sa source soit ou non explicitement mentionnée. On en retrouve la teneur fondamentale chez les païens qu'on vient de citer, mais aussi plus tard chez des chrétiens comme Boèce, Bernard de Tours (Silvestris), Dante[41], ou chez le musulman Ibn Tufayl[42] ; chez des philosophes, mais aussi chez des savants comme Ptolémée[43].

---

[39] *Timée*, 47 b 6-c 4 ; trad. Brisson, p. 144.
[40] Cicéron, *Cato maior de senectute*, XXI, 77, p. 39, 6-8.
[41] *Monarchie*, I, ix, 1.
[42] *Ḥayy ibn Yaqẓān*, p. 105 s. (arabe) et 77 s. (français), ed. Gauthier, Beyrouth, 1936. Cf. mon « Une mystique cosmologique [...] ».
[43] Préface à l'*Almageste*, p. 7, 17-24 Heiberg.

Ce que nous devons imiter n'est pas tant « la nature » que le monde céleste. Ainsi, Sénèque écrit que l'âme du sage est semblable au monde supralunaire, où il fait toujours beau (*Talis est sapientis animus, qualis mundus* [*mundi status*, v.l.] *super lunam : semper illic serenum est*)[44]. Celui-ci est le *kosmos*, le monde qui mérite son nom. L'éthique antique et médiévale contient donc une dimension selon laquelle la pratique morale doit prendre pour modèle la régularité du monde.

C'est ce qu'exprime Proclus lorsqu'il s'interroge sur les raisons de la présence, dans le *Timée*, qui est avant tout physique, d'une anthropologie. Sa dernière raison est la suivante :

> « selon l'usage des Pythagoriciens, il convenait de rattacher à l'objet contemplé le discours sur le sujet contemplant : quand en effet on sait ce qu'est le Monde, il faut, je pense, y ajouter la connaissance de ce que peut bien être le sujet qui l'examine et l'appréhende par la raison. Or, que Platon ait eu égard aussi à cet argument, il l'a prouvé quand il a dit expressément à la fin (90 d 4 s.) que, si l'on veut atteindre à la vie bienheureuse, " on doit assimiler ce qui contemple à ce qui est contemplé ". Car le Tout, lui, est éternellement heureux, et nous serons, nous aussi, heureux quand nous nous serons assimilés au Tout, car, de cette façon, nous serons remontés à notre Cause. Puisqu'en effet l'homme d'ici-bas a la même relation avec l'Univers que l'Homme Idéal avec le Vivant en Soi, et puisque là-bas les classes secondaires dépendent toujours des premières, que les parties y sont toujours inséparables des touts et sont établies en eux, quand l'homme d'ici-bas se sera assimilé à l'Univers, il imitera lui aussi son modèle sous le mode qui lui est approprié, car il sera devenu " ordonné " (κόσμιος) du fait de sa ressemblance avec l'Ordre du Monde (κόσμος), et

---

[44] Sénèque, *ad Lucilium*, [VI], 59, 16, p. 116 Reynolds.

> heureux puisqu'il se sera rendu pareil au Dieu Bienheureux »[45].

L'imitation de l'ordre du monde ne vaut pas que pour l'individu. L'ordre cosmique norme aussi la cité. Ainsi, al-Farabi se représente l'ordre de la cité idéale comme imitant la hiérarchie des sphères cosmiques :

> « alors, les parties de la cité seront reliées les unes aux autres, attachées les unes aux autres, rangées par précédence des unes et consécution des autres. Elle deviendra semblable (*Šabīh*) aux étants naturels, et ses rangs <seront> eux aussi semblables aux rangs des étants qui commencent par le premier et finissent à la matière première et aux éléments ; leur liaison et leur attachement <seront> semblables à la liaison des étants divers les uns avec les autres et à leur attachement [...]. Il faut aussi qu'il y ait, entre les parties de la nation vertueuse, une composition harmonieuse, une liaison et un concours mutuel par les actions. Il doit exister, entre les parties de la nation vertueuse et comme résultat de leurs dispositions et de leurs habitus volontaires une composition harmonieuse, une liaison, un ordre, un concours mutuel par leurs actions, semblables à ceux qui existent entre les parties du monde et résultent de leurs dispositions naturelles [...]. Et il est manifeste pour lui que le rassemblement politique, et l'ensemble qui provient dans les cités du rassemblement politique ressemble (*Šabīh*) au rassemblement des corps dans l'ensemble du monde, et il est <aussi> manifeste pour lui qu'il y a dans l'ensemble de ce que comprennent la cité et la nation des équivalents (*nẓā'ir*) de ce que comprend l'ensemble du monde. Et de même que dans le monde il y a d'abord un principe, ensuite d'autres principes selon un rang, et des étants qui viennent de ces principes, et

---

[45] Proclus, *Commentaire sur le Timée*, I, t. I, p. 5, 7-6, 6 Diehl ; trad. Festugière, t. I, p. 28-30.

> d'autres étants qui suivent ces étants selon un rang, jusqu'à ce qu'on aboutisse aux derniers des étants par le rang dans l'être, de même dans l'ensemble de ce que comprend la nation ou la cité, il y a un principe premier, ensuite d'autres principes qui le suivent, et d'autres citoyens qui suivent ces principes, et d'autres qui suivent ceux-ci, jusqu'à ce qu'on aboutisse aux derniers des citoyens par le rang dans la cité et l'humanité, si bien que se trouvent dans ce que comprend la cité des équivalents de ce que comprend l'ensemble du monde »[46].

10. POUR INTRODUIRE LA COSMONOMIE

Qui plus est, la dimension anthropologique, dans cette façon de voir, ne tient pas seulement à ce qu'une certaine image de l'homme et de son comportement est proposée par la structure de l'univers physique à titre de modèle moral. Elle est tout aussi décidément située dans la façon dont l'anthropologie est comme préparée par la physique. Celle-ci appelle l'homme à une plus pleine réalisation, par sa volonté, de ce que lui suggère déjà sa présence dans le monde.

D'une part, en effet, l'homme est lui-même conçu comme un petit monde (microcosme), comme contenant en lui, en miniature, tous les composants de l'univers physique. À chaque partie de son corps, une partie de l'univers est censée correspondre, comme le développent à satiété les auteurs, depuis l'Antiquité jusqu'au début des Temps modernes[47]. Leurs énumérations peuvent être fastidieuses et forcées, elles témoignent pourtant d'une attitude d'esprit importante : l'homme est tellement peu un étranger dans le monde qu'il est tissé de la même matière que lui. En un sens, l'imitation du monde ne fait donc que ratifier une parenté toujours déjà assurée. Il s'agira seulement d'orienter la ressemblance innée qui lie l'homme au monde vers ce que ce dernier a de plus digne d'être appelé du nom de cosmos, à savoir le monde céleste.

---

[46] Respectivement *As-Siyāsah*..., p. 84, 2 - 6 ; *Kitāb al-Millah*, p. 65, 3-6 Mahdi, tr. Mallet, p. 143 ; *Taḥṣīl as-Sa'ādah*, § 19 s. Yā Sīn / § 20 Mahdi, p. 63, 13-64, 7 Yā Sīn.

[47] Cf. Allers, « Microcosm » ; Chiesa, *Creazione*…, 68-80 ; Rico, *El pequeño mundo*[…].

Or, d'autre part, ce monde céleste n'est pas lui non plus cantonné dans le rôle d'objet attendant passivement qu'on l'imite. Il est tout ausi bien actif, en ce qu'il exerce une influence sur ce bas-monde. Bien qu'elle n'ait pas fait l'unanimité des penseurs pré-modernes, l'astrologie fut admise par une confortable majorité de ceux-ci comme un savoir légitime. C'est aussi le cas, on le sait, des pères fondateurs de l'astronomie moderne, comme par exemple Kepler. Beaucoup font des réserves devant le risque pour la liberté qu'implique le déterminisme astral. Mais dans tous les cas, cependant, on admet sans hésiter la réalité d'un certain parallélisme du ciel et de la terre, et d'une influence du premier sur la seconde[48]. De la sorte, le monde appelle et comme aimante l'homme à se laisser pénétrer par lui. Encore faudra-t-il procéder à une prudente sélection entre les bienfaits venus d'en haut et les influences supposées maléfiques.

Enfin, l'exigence d'imiter les réalités supralunaires fournit à notre être même la cause finale rendant compte de son économie : la structure même de notre corps, détaché de l'humilité de la terre et braqué vers le haut, nous permet de contempler les astres. La nature nous a ainsi redressés pour que nous puissions nous inspirer de l'exemple de ce qu'elle a produit de plus « cosmique ». L'idée se rencontre, parmi les païens, chez Cicéron, Ovide, Sénèque, Juvénal, Macrobe[49]. Elle est courante chez les Pères de l'Église[50]. Contentons-nous de citer un texte moins connu :

> « Nous ne devons rien penser de terrestre, surtout si nous savons que le dieu qui nous a fabriqués nous a faits par une disposition de son divin artifice de telle façon que, la forme droite du corps étant écartée de tout abaissement humiliant, nous ne voyons rien d'autre, dès que nous ouvrons les yeux, que le soleil, la lune, les étoiles et le domicile très beau et immortel de tous ceux-ci, à savoir le monde »[51].

On pourrait proposer pour désigner l'ensemble de cette situation le concept de **cosmonomie**. La cosmonomie n'est ni autonomie, ni

---

48 Cf. Grant, « Influence… ».
49 Respectivement : *De natura deorum*, II, lvi, 140 ; *Métamorphoses*, I, 84-86 ; *De otio, V* ; *Satires*, XV, 1436 s. ; Commentaire au *Songe de Scipion*, I, 14, 9, p. 57 Willis.
50 Cf. Pellegrino, « Status rectus… ».
51 Firmicus Maternus, *Mathesis*, VIII, 3.

hétéronomie. Elle échappe à l'alternative kantienne – comme d'ailleurs la plupart des morales concrètes. C'est en effet l'insertion dans le cosmos qui permet au sujet moral d'être authentiquement lui-même, d'être véritablement un « *autos* ». Cette conformité ne consiste donc en rien à se plier à une loi extérieure, autre (*heteros*). Pour l'homme antique et médiéval, le *kosmos* n'est pas une instance extérieure à laquelle il s'agirait d'obéir. Ou il l'est autant que la loi morale kantienne. Pour l'homme antique, « *der gestirnte Himmel über mir und das moralische Gesetz in mir* » ne se distinguent par rien d'essentiel. Pour nous êtres finis, la loi morale doit nous apparaître comme une pression extérieure, c'est-à-dire comme devoir, parce que nous avons un côté « pathologique »[52]. De la même façon, pour l'homme ancien, l'ordre du *kosmos* nous apparaît comme quelque chose d'extérieur parce que notre situation terrestre ne nous permet pas un point de vue favorable. Ce n'est que l'insertion dans le *kosmos* – comme, en style kantien, l'obéissance à la loi morale – qui nous confère une authentique liberté.

11. UNE FIN DU MONDE

Cette façon de voir le monde nous est devenue étrangère. Pour nous, il n'y a plus aucune articulation de la cosmologie et de l'éthique l'une sur l'autre, plus aucun rapport entre ce que nous savons de la structure de l'univers physique et la façon dont l'homme se pense et s'éprouve lui-même comme celui qu'il est et celui qu'il a à être. Telle est l'*opinio communis* de l'époque moderne, qui institue entre les deux domaines une séparation si radicale que la question de leur rapport ne se pose même plus. En ce sens, et pour reprendre les distinctions formulées au début de ce texte, si les progrès fantastiques de l'astrophysique ou des sciences de la terre permettent une cosmographie de plus en plus exacte et débouchent sur une cosmogonie, nous n'avons plus de cosmologie à proprement parler. Nous avons en revanche des *Weltanschauungen*, des « visions du monde » – ce que l'on pourrait se risquer à appeler, en francisant un mot déjà bien installé en castillan, des *cosmovisions*.

Dans un texte célèbre datant de 1938, Heidegger a su voir notre époque comme étant « le temps de l'image du monde »[53]. L'importance

---

[52] *Kritik der praktischen Vernunft*, § 7, p. 57.

[53] « Die Zeit des Weltbildes », repris dans *Holzwege*, Klostermann, Francfort s. Main, 1950, 69-104.

des « visions du monde » n'est qu'un aspect du processus qui fait apparaître le monde dans son ensemble comme une image. Mais il conviendrait peut-être d'apprendre à voir que la montée et le règne des « visions du monde » n'est peut-être qu'un phénomène de compensation destiné à remplacer, au sein d'une même économie intellectuelle, l'emprise désormais effacée de ce que j'ai appelé la cosmonomie.

Celle-ci constitue une époque close, que nous pouvons donc maintenant voir comme un tout. Cette période de l'histoire de la pensée humaine qui présente, du point de vue de cette articulation recherchée entre cosmologie et anthropologie, un intérêt tout particulier, s'étend de l'Antiquité post-classique à la fin du Moyen Age. Elle représente en effet une situation remarquable, celle d'une rencontre entre cosmologie et cosmographie, voire d'une quasi-coïncidence entre l'une et l'autre. Pendant une période qui a duré plus d'un millénaire, les esprits ont été en effet dominés par une cosmographie qui, d'une part, faisait l'unanimité, au moins quant à ses grands traits, et qui, d'autre part, se prêtait à l'articulation avec une éthique. Il s'agit bien d'une période, car elle a un commencement et une fin. En effet, cette liaison n'a pas toujours été le cas.

Avant cette période, plus précisément avant la victoire de la représentation de l'univers imposée par Eudoxe, Aristote, puis Ptolémée, on ne disposait pas d'une cosmographie sur laquelle on pouvait s'accorder. Chaque penseur pouvait présenter sa vision personnelle de l'univers, laquelle comportait inséparablement aussi une cosmogonie, et en défendre de son mieux la plausibilité. Avant le IVe siècle, la Grèce ne possède pas de vision unifiée de l'univers physique[54]. Le triomphe subséquent d'une cosmographie unique élimina ou priva de pertinence toute tentative de cosmogonie, par exemple en la reléguant dans le domaine mythique (ainsi, dans le *Timée* de Platon). C'est à l'intérieur d'un même cadre que les cosmologies purent s'affronter. Ou plutôt, le problème cosmologique pouvait désormais se poser. On savait face à quoi, ou plutôt dans quoi il fallait se situer et chercher à résoudre le problème de l'homme.

Par la suite, en revanche, soit en gros avec l'époque moderne, la cosmographie qui pouvait faire l'accord des esprits ne se prêtait plus à une surdétermination éthique, négative ou positive. Le cosmos moderne est éthiquement indifférent. L'image du monde qui sort de la physique

---

[54] Cf. G.E.R. Lloyd, « Greek Cosmologies ».

newtonienne est celle d'un jeu de forces aveugles, où il n'y a pas de place pour la considération du Bien.

Pour que la morale puisse se détacher de la cosmonomie et être conçue exclusivement comme « autonomie », il faudra que le cosmos ait déjà perdu sa fonction constitutive par rapport au sujet humain, et qu'il n'apparaisse plus que comme le décor indifférent où se déploie une activité humaine qui lui est au fond étrangère, et qui ne lui doit rien de ce qui la fait accéder à son humanité. Or, c'est exactement ce qui est arrivé au concept de nature, avec les temps modernes. L'idée d'une imitation morale de la nature est devenue impossible, parce que notre concept de nature s'est modifié. Le monde ne peut plus nous aider à devenir des hommes.

12. LA NOUVELLE IMAGE DU MONDE

Les Romains et les Chinois avaient observé dès l'Antiquité l'apparition d'étoiles nouvelles (*novae*), mais ils n'avaient pas voulu ou pas pu les « voir ». L'observation de deux de celles-ci, en 1572 (Tycho-Brahé), puis 1604 (Kepler), désormais devenue pertinente, montre que les cieux ne sont pas immuables. Les philosophes en ont vu aussitôt les conséquences[55]. La découverte des taches du soleil par Christoph Scheiner, s.j. (1611), une véritable nouveauté celle-là, car seuls les nouveaux instruments optiques la rendaient possible, montre que les cieux et la terre sont faits de la même matière, impure et corrompue. Galilée abandonne la division aristotélicienne entre sublunaire et supralunaire[56]. Plutôt que d'illustrer ces révolutions astronomiques par des textes de l'époque, citons ici Salomon Maïmon, qui les mentionne encore comme des nouveautés dans un texte publié aussi tard que 1791 :

> « Il faut que tu saches, lecteur éclairé, qu'en vérité, conformément à ce qui a été démontré chez les physiciens et les astronomes modernes, les individus célestes, eux aussi, changent comme les individus terrestres que le Maître [...] mentionne, du fait qu'en vérité la matière des corps célestes n'est pas un cinquième corps, comme le pensait

---

[55] Cf. Juste-Lipse, *De constantia* (1572), I, 16 ; commodément accessible et traduit dans J. Lagrée, *Juste Lipse*. La restauration du stoïcisme. Étude et traductions de divers traités stoïciens, Vrin, Paris, 1994, p. 132 s.

[56] *Dialogo*, I ; *Opere*, t. VII, p. 62 ss.

> Aristote, mais elle est la même matière que la terre. Et non seulement les modernes ont trouvé des planètes nouvelles qui n'étaient pas encore connues à l'époque du Maître (cinq satellites de Jupiter et quatre satellites de Saturne), mais, récemment, une étoile nouvelle a été révélée aux modernes (le cinquième satellite de Jupiter). De même, les savants dont on a parlé ont découvert des étoiles fixes nouvelles que les anciens n'avaient pas comptées, et aussi d'autres étoiles qui étaient connues des anciens et qui ont disparu de nos yeux au point que l'on ne sait plus où se trouvait leur place »[57].

Le concept de monde est placé par la physique moderne, post-galiléenne, dans une situation nouvelle. Il reçoit d'une part une pertinence nouvelle. Kepler intitule un traité *Mysterium cosmographicum* (1595). Descartes écrit (sans le publier) un *Traité du monde* (1633). L'idée d'ordre est déterminante : monde s'oppose à chaos[58]. Pour Leibniz, un monde – plusieurs sont possibles, parmi lesquels Dieu choisit le meilleur – est un assemblage de compossibles. Mais le concept de monde est-il pensé comme tel ? Descartes emploie, peut-être sans le remarquer, l'expression « seul dans le monde » au moment où il opère pourtant la remise en doute la plus radicale de l'existence des choses extérieures[59]. Tout se passe comme si le monde était l'horizon de présence que suppose même toute absence. D'un autre côté, le concept de monde passe au pluriel pour désigner des terres habitées ou des systèmes analogues à celui du soleil. Mais ces mondes ne composent plus un « uni-vers ». Peut-on alors encore parler de cosmologie ? Pour Jacques Merleau-Ponty, l'Occident a cessé d'en avoir une avec la fin du monde aristotélico-ptolémaïque, entraînée par Copernic, Galilée, puis Newton. Les « mondes » ne formaient plus alors un tout. Nous avons recommencé à en avoir une avec Eddington, depuis que nous avons un modèle unifié, désormais dynamique, de l'unité du cosmos[60].

---

[57] *Giv'at ha-Moreh*, sur Guide, I, 11, p. 46 Bergman-Rotenstreich. Cf. également, dans le commentaire à I, 15, p. 52, la prise en compte de l'illégitimité de la distinction entre le lourd et le léger.

[58] Descartes, *Traité du monde*, VI, AT XI, p. 34 s.

[59] *Meditationes*, II, AT VII, p. 25 ; III, p. 42 s.

[60] J. Merleau-Ponty & B. Morando, *Les trois étapes*[…].

Quant à la pertinence anthropologique et éthique de la cosmologie, elle est minée. La physique newtonienne (1688) montre que le ciel est régi par les mêmes lois que ce bas monde ; la lune et la pomme leur sont soumises également. Et ces lois qui ne sont pas précisément celles d'une conduite morale exemplaire, puisque, en dernière instance, seule la force y compte. Dans le cosmos newtonien, les gros astres attirent les petits. Tout s'y passe de la même façon que, dans le monde spinoziste, « les gros poissons mangent les petits ». Si les corps célestes ne se mangent pas les uns les autres, c'est uniquement parce que leur vitesse leur permet d'échapper à l'attraction qu'ils exercent les uns envers les autres, rien de plus. Les « gros » se grossiraient encore de la masse des « petits », si ceux-ci ne parvenaient à leur échapper. On peut en déduire que le même combat qui oppose sur la terre les espèces animales et les individus est aussi le ressort du mécanisme céleste. Schopenhauer a tiré cette conclusion, là où il mentionne :

> « die beständige Spannung zwischen Zentripetal- und Zentrifugalkraft [...], welche das Weltgebaüde in Bewegung erhält und selbst schon ein Ausdruck ist jenes allgemeinen, der Erscheinung des Willens wesentlichen Kampfes [...] »[61].

13. LA CONSCIENCE POST-COPERNICIENNE

La nouvelle astronomie, après Copernic et ses successeurs, a eu des conséquences sur la vision du monde moderne. Vu à la moderne, l'ensemble des réalités physiques ne pouvait plus porter le nom de monde. Le passage « du monde clos à l'univers infini » (Koyré) est aussi reflété par une évolution dans le vocabulaire, qui aboutit à l'état de choses signalé plus haut :

> « The long use of *word* to mean an object so patterned and unified as the geocentric *kosmos* went so far as to disqualify it as a name for what we now call " universe ". When once the old model is broken, we become aware that a detailed knowledge of the Region of all regions will be forever unattainable. We now need a word not for

[61] *Die Welt als Wille und Vorstellung*, II, 1, § 27, p. 220.

some specific and imaginable object but for " whatever the totality may be ", and *universe* is such a word. Furthermore, the same process which broke the old model threw a new meaning on *world* <celui de " domaine habité "> which, while it lasted, made it impossible, without grave inconveniences, to call the totality the *world* »[62].

Quant aux conséquences sur la façon dont l'homme perçoit sa présence dans le monde, elles ne consistent pas, comme on le croit trop souvent, en ce que l'homme, cessant de se sentir au centre du monde, en aurait été humilié. La place centrale, dans la cosmologie ptolémaïque, était plutôt un lieu de honte qu'une place d'honneur, et la terre, une poubelle plutôt qu'un trône[63]. Ce qui frappe les esprits (*certains* esprits) est plutôt l'absence de point de référence due à un univers supposé désormais infini. La fin du cosmos aristotélicien fournit à John Donne, dans un texte peut-être trop cité, tout un arsenal d'images :

« The new philosophy calls all in doubt,
The element of fire is quite put out ;
The sun is lost and the earth, and no man' wit
Can well direct him where to look for it.
And freely men confess that this world's spent,
When in the planets and the firmament
They seek so many new ; they see that this
Is crumbled out again to his atomies.
'Tis all in pieces, all coherence gone ;
All just supply, and all relation »[64].

Le libertin mis en scène par Pascal – et surtout pas, selon un contresens répandu, Pascal lui-même – s'effrayait déjà devant « le silence éternel de ces espaces infinis »[65]. Adam Smith exprime par la même image le cauchemar d'un monde sans Dieu :

« To this universal benevolence, on the contrary, the very suspicion of a fatherless world, must be the most melancholy of all reflections ; from the thought that all the unknown regions of infinite and incomprehensible space may be filled

---

[62] C.S. Lewis, *Studies*..., p. 251.
[63] Cf. mon « Geozentrismus... ».
[64] John Donne, *Anatomy of the World*, First Anniversary (1611).
[65] *Pensées*, Br. 206/Lafuma 201 et cf. Br. 194/Lafuma 427, début.

with nothing but endless misery and wretchedness. All the splendour of the highest prosperity can never enlighten the gloom with which so dreadful an idea must necessarily over-shadow the imagination »[66].

Dans un texte écrit vers la fin de sa vie, Gérard de Nerval (1808-1855) rêve que la terre s'est détachée :

« Les étoiles brillaient dans le firmament. Tout à coup il me sembla qu'elles venaient de s'éteindre à la fois comme les bougies que j'avais vues à l'église. Je crus que les temps étaient accomplis et que nous touchions à la fin du monde annoncée dans l'*Apocalypse* de saint Jean. Je croyais voir un soleil noir dans le ciel désert et un globe rouge de sang au-dessus des Tuileries. Je me dis : " La nuit éternelle commence, et elle va être terrible. Que va-t-il arriver quand les hommes s'apercevront qu'il n'y a plus de soleil ? " Je revins par la rue Saint-Honoré, et je plaignais les paysans attardés que je rencontrais. Arrivé vers le Louvre, je marchai jusqu'à la place, et là un spectacle étrange m'attendait. À travers des nuages rapidement chassés par le vent, je vis plusieurs lunes qui passaient avec une grande rapidité. Je pensai que la terre était sortie de son orbite et qu'elle errait dans le firmament comme un vaisseau démâté, se rapprochant ou s'éloignant des étoiles qui grandissaient ou diminuaient tour à tour. Pendant deux ou trois heures je contemplai ce désordre [...] »[67].

Les images sont de toutes provenances : une source est indiquée, l'*Apocalypse*. D'autres sont citées sans leur provenance, mais il n'est pas difficile de trouver là la rencontre de plusieurs registres : le soleil noir, déjà évoqué dans un poème antérieur comme celui de la Mélancolie. Mais ici, le noir est le signe d'un vide, non d'un rayonnement négatif. Le soleil n'est pas seulement caché, il n'est plus là, d'où la conséquence : la

[66] *Theory of Moral Sentiments*, Liberty Classics, Indianapolis, p. 235.
[67] Nerval, *Aurelia*, IIe partie, iv, p. 335 s. Folio.

terre est détachée de son orbite, et erre sans but. Le ciel est comparé à la mer, et l'errance à celle d'un vaisseau démâté, image qui restera classique elle aussi[68]. Le passage s'achève sur le paradoxe implicite : « contempler le désordre ». La contemplation antique tirait son sens de ce qu'elle vise un ordre sur lequel le sujet qui contemple se modèle. Ici, la situation qui résulte est digne d'intérêt précisément parce qu'elle est désordonnée.

Nietzsche utilise lui aussi des images astronomiques. Notamment pour exprimer l'idée de la « mort de Dieu », dans le célèbre aphorisme « Der tolle Mensch » :

> « Wie vermochten wir das Meer auszutrinken ? Wer gab uns den Schwamm, um den ganzen Horizont wegzuwischen ? Was taten wir, als wir die Erde von ihrer Sonne losketteten ? Wohin bewegt sie sich nun ? Wohin bewegen wir uns ? Fort von allen Sonnen ? Stürzen wir nicht fortwährend ? Und rückwärts, seitwärts, vorwärts, nach allen Seiten ? Giebt es noch ein Oben und ein Unten ? Irren wir nicht durch ein unendliches Nichts ? Haucht uns nicht der leere Raum an ? Ist es nicht kälter geworden ? Kommt nicht immerfort die Nacht und mehr Nacht ? »[69].

Comme chez Nerval, il s'agit du retard par rapport au fait de la conscience qu'on en prend. Ailleurs, Nietzsche se réfère explicitement à la nouvelle astronomie : « Seit Copernicus rollt der Mensch aus dem Centrum ins X »[70]. Le tournant copernicien reçoit ici, il vaut de le noter, une interprétation tout à fait paradoxale, puisque la cosmologie héliocentrique enchaîne plutôt résolument la terre au soleil.

14. L'IMPOSSIBILITÉ D'UNE IMITATION DU MONDE

Le rapport antico-médiéval au monde devient impossible. On pourrait, certes, « imiter la nature ». Mais cela aurait des conséquences amorales, voire immorales. Nietzsche l'a expliqué dans un passage célèbre où il s'en prend à l'idéal stoïcien du « vivre selon la nature »[71]. Mais l'idée

---

[68] Rimbaud, « Le bateau ivre » ; Mallarmé, « Brise marine ».
[69] Nietzsche, *Die fröhliche Wissenschaft*, III, § 125 ; KSA, t. 3, p. 481.
[70] Nietzsche, fgt. automne 1885-automne 1886, 2 [127] ; KSA, t. 12, p. 127.
[71] *Jenseits von Gut und Böse*, § 9, KSA, t. 5, p. 21 s.

vient du XVIIIe siècle. Ainsi, Sade fonde la nécessité, voire la légitimité de la destruction, bref, de ce que l'on a appelé plus tard le « sadisme », sur le fait que la nature se renouvelle en détruisant :

> « Ce n'est que par des forfaits que la nature se maintient et reconquiert les droits que lui enlève la vertu. Nous lui obéissons donc en nous livrant au mal ; notre résistance est le seul crime qu'elle ne doive jamais nous pardonner »[72].

Quelques décennies plus tard, mais dans une pièce dont la date dramatique est proche de l'époque où Sade a écrit ses œuvres, Georg Büchner place dans la bouche de Saint-Just un discours à la Convention dans lequel il justifie la Terreur par l'exemple de la nature :

> « Es scheint in dieser Versammlung einige empfindliche Ohren zu geben, die das Wort " Blut " nicht wohl vertragen können. Einige allgemeine Betrachtungen mögen sie überzeugen, daß wir nicht grausamer sind als die Natur und die Zeit. Die Natur folgt ruhig und unwiderstehlich ihren Gesetzen ; der Mensch wird vernichtet, wo er mit ihnen in Konflikt kommt. Eine Änderung in den Bestandteilen der Luft, ein Auflodern des tellurischen Feuers, ein Schwanken in dem Gleichgewicht einer Wassermasse und eine Seuche, ein vulkanischer Ausbruch, eine Überschwemmung begraben Tausende. Was ist das Resultat ? Eine unbedeutende, im großen Ganzen kaum bemerkbare Veränderung der physischen Natur, die fast spurlos vorübergegangen sein würde, wenn nicht Leichen auf ihrem Wege lägen. Ich frage nun : soll die geistige Natur in ihren Revolutionen mehr Rücksicht nehmen als die physische ? Soll eine Idee nicht ebensogut wie ein Gesetz der Physik vernichten dürfen, was sich ihr widersetzt ? Soll überhaupt ein Ereignis, was die ganze Gestaltung der moralischen Natur, das heißt der Menschheit, umändert, nicht durch Blut gehen dürfen ? Der Weltgeist bedient sich in der geistigen

---

[72] *Juliette*, fin.

> Sphäre unserer Arme ebenso, wie er in der physischen Vulkane und Wasserfluten gebraucht. Was legt daran, ob sie an einer Seuche oder an der Revolution sterben ? »[73].

Nous ne pouvons plus prendre quoi que ce soit dans le monde comme exemple d'action morale. Prenons comme exemple le distique célèbre de Schiller, « Das Höchste » :

> « Suchst du das Höchste, das Größte ? Die Pflanze kann es dich lehren.
> Was sie willenlos ist, sey du es wollend – das ists ! »[74].

Une imitation matérielle des plantes serait scandaleuse : les fleurs ne sont-elles pas des organes sexuels exposés en plein jour au centre d'une jupe bariolée, impudiquement retroussée ? L'imitation ne peut être que formelle. Ce n'est pas l'activité de la plante, mais uniquement sa beauté que l'on peut imiter. De plus, il faut pour ce faire oublier à quoi cette beauté sert au juste, par exemple à attirer les insectes qui transporteront le pollen. Il faut oublier, en un mot, que la plante n'est rien moins que « *willenlos* ». En conséquence, il n'y a plus d'imitation morale de la nature. Le Beau n'est, dans le meilleur des cas, qu'une invitation subjective à la moralité, et en aucun cas l'expression d'un Bien avec lequel il se laisserait « convertir », au sens de la doctrine scolastique des transcendantaux.

15. Le mal sans appel

Les exemples que Sade ou le Saint-Just de Büchner citent en faveur de leur théorie (volcans, tremblements de terre, raz de marées, etc.) étaient connus depuis toujours. Le monde antique et médiéval n'ignorait pas le mal qui règne sur terre. Ainsi, les Anciens savaient très bien que les animaux se conduisent les uns envers les autres selon de purs rapports de force. Hérodote, par exemple, savait que l'équilibre des espèces animales était assuré par la fécondité plus ou moins grande des animaux par rapport à leur capacité prédatrice :

---

[73] *Dantons Tod* (hiver 1835), acte II, scène 7, p. 43 Reclam.
[74] « Die Horen » (1795-1796), dans *Werke*, Nationalausgabe, 1. Band, Gedichte, Weimar, 1943, p. 259.

> « la providence du divin (τοῦ θείου ἡ προνοίη), comme on doit s'y attendre, étant habile (σοφή), quant aux animaux qui sont d'âme lâche et mangeables, elle les a tous faits prolifiques, afin qu'ils ne disparaissent pas dévorés par les autres, et ceux qui sont dangereux et cruels, elle les a tous faits peu féconds »[75].

Ainsi, le lapin est fécond. La lionne, en revanche, n'enfante qu'une fois, le lionceau ne pouvant naître, croyait-on, qu'en déchirant la matrice de ses ongles[76]. Lucrèce sait que chaque être vivant dispose d'une arme ou d'une défense quelconque, faute de quoi il ne pourrait survivre[77].

Les Anciens ne se faisaient pas non plus d'illusions sur la bonté naturelle de l'homme : dans le dialogue entre les Athéniens et les Méliens, Thucydide nous livre une vision d'un machiavélisme radical[78]. Ils n'ignoraient pas que les gros poissons mangent les petits. Cet exemple était même pour eux proverbial, ainsi que l'exprime l'idée indienne de « loi des poissons » (*matsyanyaya*)[79]. Mais l'observation de la conduite animale servait pour eux de repoussoir à l'éthique humaine. L'exemple des poissons est pour Hésiode le modèle de ce que l'homme **ne** doit **pas** faire[80]. Nemesius d'Emèse, dans un passage pour lequel les éditeurs ne mentionnent pas de sources, nuance en rappelant que la chaîne alimentaire a une fin : certains poissons, les plus petits, qu'il appelle φυκία, mangent ce qu'on pourrait appeler l'herbe de la mer[81]. Les rabbins du Talmud, de leur côté, évoquent le risque d'une retombée dans la loi de la jungle pour justifier la nécessité du gouvernement. Sans loi, « les hommes s'avaleraient vifs les uns les autres »[82]. Ils ne se contenteraient pas par là de transgresser le commandement qui interdit le meurtre, mais s'abstiendraient même de respecter la plus élémentaire des

---

[75] Cf. Hérodote, III, 108, 2.
[76] Malgré les sarcasmes d'Aristote (*Histoire des Animaux*, VII, 31, 579 b 2), la légende se trouve encore, par exemple chez s. Basile, *In Hexaemeron*, IX, 5, p. 502 s. Giet (SC, 26 bis).
[77] *De natura rerum*, V, 871 - 877.
[78] Cf. Thucydide, V, 85 - 111.
[79] Cf. R. Lingat, *The Classical Law of India*, New Delhi/Berkeley, 1973, p. 207 s.
[80] Cf. Hésiode, *Les travaux et les jours*, 277 s.
[81] Nemesius, *De natura hominis*, § 1, p. 12, 17-13, 1.
[82] *Pirqey Aboth*, III, 2.

lois noachiques, celle qui interdit « le membre <arraché> à un vivant »[83], par laquelle l'homme se distingue de l'animal qui se jette sur sa proie.

Mais un gouffre sépare ces textes des textes modernes qu'on vient de citer : les Anciens pouvaient faire appel, et renvoyer au supralunaire, pour montrer que, dans l'écrasante majorité des cas, la nature est gouvernée par le Bien. Pour nous, en revanche, ce chemin est devenu une impasse. Nous ne connaissons plus la distinction entre les domaines sublunaire et supralunaire. Pour nous, l'univers est une totalité, une tunique sans couture. Nous ne sommes plus en droit de faire la différence entre des niveaux du monde qui auraient une plus ou moins grande valeur. C'est ce qu'a perçu un moraliste anglais de la fin du siècle dernier. Au cours d'une argumentation qui vise à restreindre la qualité de « bon » ou de « mauvais » à l'humain, Henry Sidgwick écrit :

> « No doubt there is a point of view, sometimes adopted with great earnestness, from which the whole universe and not merely a certain condition of rational or sentient beings is contemplated as ‘ very good ’ : just as the Creator in *Genesis* is described as contemplating it. But such a view can scarcely be developed into a method of Ethics. For practical purposes, we require to conceive some parts of the universe as at least less good than they might be. And we do not seem to have any ground for drawing such a distinction between different portions of the non-sentient universe, considered in themselves and out of relation to conscious or sentient beings »[84].

Ce qui, pour Sidgwick, était une évidence qu'il n'y avait pas besoin de fonder, à savoir la conviction, exprimée à la fin du texte, qu'aucune partie du monde n'est « meilleure » qu'une autre, aurait été un scandale pour l'antiquité et le Moyen Age. Pour arriver à cette prétendue évidence, il a fallu les efforts des savants pour détruire l'évidence précédente, qui était contraire. Il en est ainsi à la suite d'un mouvement enclenché dès l'aube des temps modernes.

---

[83] *Exode* 20, 13 ; *Deutéronome* 5, 17, puis *Genèse* 9, 4 ; b*Hullin* 101 b.
[84] *The Methods of Ethics*, Londres, 1884 (3e éd.), I, ix, § 4, p. 109 s.

16. UNIVERSALITÉ DU MAL

En conséquence, le monde est tel que le mal existe en chacune de ses parties, sur la terre comme au ciel, et pas moins au ciel que sur la terre. Ainsi, Leopardi peut écrire : « il male è cosa commune a tutti i pianeti dell'universo »[85].

Le monde apparaît de plus en plus comme le champ de bataille où s'affrontent des puissances aveugles. On trouve cette façon de voir chez les philosophes modernes. Ainsi Pic de la Mirandole, pour lequel la philosophie naturelle (physique) ne peut enseigner qu'une paix relative, puisque son principe est héraclitéen : la Guerre père de toutes choses[86]. De même, Shlomo Ibn Verga (début du XVIe) fait dire au pape, censé justifier l'expulsion des juifs d'Espagne :

> « Il est bien connu que nous sommes composés de quatre éléments et que la nature des <corps> simples est <présente> dans ce qui est composé. Nous constatons que chaque élément <s'occupe> de dominer et cherche à absorber l'autre et à le transformer en sa propre substance ; ainsi, nous constatons que l'eau brûle toute chose si l'eau ne l'en empêche pas [...]. Il en est de même pour les animaux, oiseaux et poissons : le plus fort avale l'autre. De même pour les peuples : la nation dominante cherche à transformer les autres en sa propre substance. Et si j'avais la possibilité que personne d'autre au monde ne règne que moi, j'y consacrerais toute ma capacité et tout mon pouvoir »[87].

La même optique se retrouve dans toute la littérature moderne, et d'autant plus déchaînée qu'elle n'est plus contrebalancée par rien : au XVIIIe, on la trouve chez Goethe, dans son premier roman, le *Werther*. À l'occasion d'une saute d'humeur due aux malheurs sentimentaux du jeune homme, la nature apparaît sous un tout autre jour. Elle semblait bourdonnante de fécondité, comme animée par un esprit qui ne cesse de créer et se réjouit de tout ce qui vit, et comme offrant à celui qui le

---

[85] « Dialogo della terra e della luna » (24-28.4.1824), dans *Operette morali*, ed. P. Ruffilli, Garzanti, s.l., 1984, p. 79.

[86] *Oratio de hominis dignitate*, p. 22 Boulnois-Tognon.

[87] *Shevet Yehuda*, ed. M. Wiener, Hannovre, 1924, p. 79.

contemple la possibilité de s'unir à la joie du Créateur. Elle laissait deviner un dieu stoïcien ou spinoziste. Puis un rideau se lève, et le théâtre de la vie infinie se change en l'abîme du tombeau éternellement ouvert. Le jeune Werther conclut :

> « Da ist kein Augenblick, der nicht dich verzehrte und die Deinigen um dich her, kein Augenblick, da du nicht ein Zerstörer bist, sein mußt ; der harmloseste Spaziergang kostet tausend armen Würmchen das Leben, es zerrüttet ein Fußtritt die mühseligen Gebäude der Ameisen, und stampt eine kleine Welt in ein schmähliches Grab. Ha ! Nicht die große seltene Not der Welt, diese Fluten, die eure Dörfer wegspülen, diese Erdbeben, die eure Städte verschlingen, rühren mich ; mir untergräbt das Herz die verzehrende Kraft, die in dem All der Natur verborgen liegt ; die nichts gebildet hat, das nicht seinen Nachbar, nicht sich selbst zerstörte. Und so täumle ich beängstigt ! Himmel und Erde und ihre webenden Kräfte um mich her ! Ich sehe nichts als ein ewig verschlingendes, ewig wiederkäuendes Ungeheuer »[88].

L'idée fondamentale est assez claire : voir le monde comme beau et bon relève de l'illusion ou de la myopie ; sa réalité profonde est terrible – opposition que l'on verra réapparaître plus récemment chez Hugo von Hofmannsthal, avec le même retournement que dans le *Werther*[89], ou encore chez Céline[90], Carl Zuckmayer[91], ou Dino Buzzatti[92]. Mais on notera quelques détails de grande conséquence : d'une part, ciel et terre sont placés sur le même plan, sans que la monstruosité de la nature, nouveau *Kronos* dévorant ses enfants, soit réservée à ce bas-monde ; d'autre part, ce n'est pas seulement ce qui, dans la nature, détruit l'homme de façon spectaculaire et soudaine qui

---

[88] *Werther*, Livre I, lettre du 18 août.
[89] *Der Jüngling und die Spinne* .
[90] *Voyage au bout de la nuit*, scènes d'Afrique.
[91] *Die Geschichte vom Tümpel* (1926), dans *Gedichte, Erzählungen* (GW 1), Fischer, Francfort, 1960, 223 - 230.
[92] « Dolce notte », dans *Il colombre* [...], Mondadori, 1966, 193-199 ; tr.fr. « Douce nuit », dans *Le K.*, Laffont, Paris, 1967, 160-165.

suscite l'horreur. Goethe pense certainement, l'allusion est trop claire, au fameux tremblement de terre de Lisbonne. Celui-ci, vieux d'à peine vingt ans (1755) avait frappé d'horreur l'Europe des Lumières et avait suscité la réflexion de bien des têtes pensantes, de Voltaire à Kant. Devant ce genre de catastrophes, on pouvait affirmer sa dignité d'être pensant sur le fond de l'indifférence cosmique. Face à la nature qui le détruit, l'homme garde sa noblesse, car lui seul sait qu'il meurt : « Mais quand l'univers l'écraserait, l'homme serait encore plus noble que ce qui le tue, puisqu'il sait qu'il meurt et l'avantage que l'univers a sur lui, l'univers n'en sait rien »[93]. Or, dans le texte du *Werther*, il s'avère que nous sommes nous-mêmes parties prenantes de la destruction : notre existence même suppose la mort d'autres êtres vivants, nous ne pouvons faire un pas sans que cela coûte la vie à des millions d'animalcules. Nous sommes à ces bestioles ce que le tremblement de terre est à l'homme. Nous sommes une partie de la nature et, comme tels, il nous faut jouer le jeu. Le ciel et la terre, mais aussi l'homme et la nature, sont logés à la même enseigne.

17. UNE MORALE EXTRA-MORALE

Puisque l'homme ne fait pas exception, ce qui le distingue du reste du monde est pensé sur le modèle du monde lui-même. La morale elle-même est vue d'une façon « extra-morale (*außermoralisch*) », pour parler comme Nietzsche. La tendance est séculaire, à partir des penseurs du libéralisme, et l'on pourrait multiplier les exemples. J'en citerai un seul, la parabole des porcs-épics chez Schopenhauer : ils se rapprochent les uns des autres pour se tenir au chaud, mais doivent alors se piquer, jusqu'à ce qu'ils trouvent la distance optimale qui leur permet de ne pas avoir trop froid, tout en ne se faisant pas trop mal[94]. La comparaison me semble intéressante en ce qu'elle confirme sans le dire le rôle directeur de la cosmologie pour la vision d'ensemble du monde. Tout se passe en effet comme si elle transposait dans la zoologie une situation déjà observée en mécanique céleste : la morale n'a rien de moral, elle ne s'occupe ni du bien ni du mal. Il y va seulement d'une certaine distance, d'un certain équilibre. Le problème est de trouver le point où les forces contraires s'annulent, comme dans le cosmos newtonien.

---

93 Pascal, *Pensées*, Br. 347 ; Lafuma 200.
94 *Paralipomena*, ch. 31, § 396.

Un passage de Nietzsche me semble bien résumer la sensibilité moderne quant au monde :

> « Hüten wir uns, zu denken, daß die Welt ein lebendiges Wesen sei. Wohin sollte sie sich ausdehnen ? Wovon sollte sie sich nähren ? Wie könnte sie wachsen und sich vermehren ? Wir wissen ja ungefähr, was das Organische ist : und wir sollten das unsäglich Abgeleitete, Späte, Seltene, Zufällige, das wir nur auf der Krüste der Erde wahrnehmen, zum Wesentlichen, Allgemeinen, Ewigen umdeuten, wie es Jene thun, die das All ein Organismus nennen ? Davor ekelt mir. Hüten wir uns schon davor, zu glauben, daß das All eine Maschine sei ; es ist gewiss nicht auf ein Ziel construirt, wir thun ihm mit dem Wort " Maschine " eine viel zu hohe Ehre an. Hüten wir uns, etwas so Formvolles, wie die kyklischen Bewegungen unserer Nachbar-Sterne überhaupt vorauszusetzen ; schon ein Blick in die Milchstrasse lässt Zweifel auftauchen, ob es dort nicht viel rohere und widersprechendere Bewegungen giebt, ebenfalls Sterne mit ewigen geradlinigen Fallbahnen und dergleichen. Die astrale Ordnung, in der wir leben, ist eine Ausnahme ; diese Ordnung und die ziemliche Dauer, welche durch sie bedingt ist, hat wieder die Ausnahme der Ausnahmen ermöglicht : die Bildung des Organischen. Der Gesammt-Charakter der Welt ist dagegen in alle Ewigkeit Chaos, nicht im Sinne der fehlenden Notwendigkeit, sondern der fehlenden Ordnung, Gliederung, Form, Schönheit, Weisheit, und wie alle unsere ästhetischen Menschlichkeiten heissen [...] »[95].

Nietzsche détruit successivement les principales métaphores qui permettent de penser le monde : organisme, machinerie, jusqu'à l'idée même d'ordre. Nous sommes de la sorte placés aux antipodes de la vision grecque du monde comme d'un ordre (κόσμος) s'installant dans la

---

[95] Nietzsche, *Die fröhliche Wissenschaft*, III, § 109 ; KSA, t. 3, p. 467 s.

béance d'un vide originel (χάος). Le monde n'est plus *kosmos*, mais justement, chaos. Il n'y a pas là, peut-être à l'encontre même de certaines formules de Nietzsche, un jugement scientifique que pourrait réfuter la découverte de nouvelles lois et de régularités plus fines – découverte qui n'a pas cessé depuis Nietzsche. Il s'agit de la conséquence du découplage par rapport à l'homme, du caractère radicalement in-humain, parce que a-moral, du monde.

Antiquité et Moyen Age avaient été dominés par le modèle platonicien. Son recul invite à reconsidérer les modèles qu'il a supplantés au moment où il restait maître du champ de bataille, ou qu'il a obligés à coexister avec lui en les influençant. On peut en décrire trois.

18. L'ALTERNATIVE ÉPICURIENNE

On peut interpréter l'épicurisme comme une révocation de la pertinence de l'idée de monde. L'étude de la physique a bien une fin morale, mais le but est atteint indirectement. À la différence de ce qu'on a vu plus haut[96], ce n'est pas en s'inspirant du monde connu, mais de la connaissance du monde, que l'on peut obtenir une réforme morale[97]. L'effet de la physique n'est pas objectif, mais purement subjectif. L'univers physique n'est pas intéressant en soi, au sens fort de cet adjectif : nous n'avons pas à passer par lui pour atteindre ce que nous avons à être. La structure de l'univers est en dernière instance indifférente à l'existence humaine et à son libre épanouissement. Il n'est pas question d'imiter des phénomènes physiques, comme chez Platon, mais de s'en distancier. La connaissance n'a pas pour but l'assimilation, mais l'objectivation, comme moyen de dissimilation. Le bonheur n'est en rien favorisé par l'examen des phénomènes célestes, voire l'excès de celui-ci peut le compromettre[98]. Les théories que l'on en proposera n'auront que le mérite de supprimer l'angoisse devant ce qui, dans l'univers, pourrait la susciter[99] ; elles pourront donc être plusieurs. La mise au pluriel des mondes interdit de considérer notre monde, manière fortuite et passagère parmi d'autres dont des atomes se sont associés[100], comme un modèle. Qu'il ne soit qu'un exemplaire l'empêche d'être un exemple. De même, la fragilité,

---

[96] Cf. supra, § 8, p. 40-42.
[97] Cf. Torquatus dans Cicéron, *De finibus*, I, xix, 64, p. 28 Schiche.
[98] Lucrèce, V, 1204-1240.
[99] *Lettre à Pythoclès*, § 85 ; *Kuriai Doxai*, XI.
[100] *Lettre à Hérodote*, § 73.

voire l'imminence de sa dissolution selon Lucrèce[101] interdit qu'on y cherche un appui ou un recours. Enfin, Epicure, selon plusieurs témoignages indirects, aurait situé les dieux dans les « intermondes ». Les dieux étant ainsi radicalement non-cosmiques, les imiter cessera de coïncider avec imiter le monde.

On peut trouver des équivalents modernes de cette attitude. C'est le cas de cette tendance qui préfère éviter de parler de la vérité comme de l'objet de la philosophie pour lui assigner un rôle avant tout thérapeutique. Elle est représentée aujourd'hui par Rorty : l'accord entre les hommes, ce qu'Epicure appelait l'« amitié » (φιλία), auquel Rorty préfère donner le nom moderne de « démocratie », est plus important que la vérité inaccessible. On peut y rattacher aussi, au niveau de l'attitude d'ensemble, un certain « épicurisme » dans lequel le sens populaire de l'adjectif est, pour une fois, relativement légitime. Le quatrième Évangile résumait le mode authentique de la présence dans le monde sous le chef de la « tribulation » (θλῖψις) : « dans le monde, vous avez de la tribulation » (*Jean* 16, 33). Celle-ci est désormais remplacée par le « confort » – un confort qui a lui-même changé de sens, par rapport à ce qu'il était encore dans le fameux texte où Thomas More le faisait dialoguer, justement, avec la tribulation. Ce confort n'est plus ce qui, nourrissant le courage, permet d'endurer la tribulation comme telle ; il est désormais supposé s'y substituer.

19. L'ALTERNATIVE GNOSTIQUE

Le monde a été fait par un démiurge qui n'est pas le Dieu suprême, mais un être qui lui est subordonné, et qui est ignorant ou mauvais. Le monde est beau, mais il n'est pas bon. Le monde est même la plénitude du mal[102]. Soit parce qu'il est bâclé, soit, pour certains, parce qu'il est méchant, malveillant. Il est bien d'un ordre rigoureux et incontestable, voire, d'une beauté fascinante, mais son ordre sert en fait à rendre impossible l'évasion de l'âme. Et sa beauté est un piège destiné à la séduire. Une puissance mauvaise règne au firmament – c'est-à-dire justement là où le Bien devrait être à son comble. Le monde est une prison pour l'âme, un lieu d'angoisse[103]. L'âme est abandonnée dans le

---

[101] V, 95-106, Cf. II,

[102] *Hermès Trismégiste*, VI, 4, t. 1, p. 74 Nock-Festugière.

[103] Basilide, dans Clément, *Stromates*, IV, 12 ; 88, 5 [p. 287, § - 8 Stählin].

monde comme un avorton[104], ou jetée en celui-ci par un mouvement violent[105]. Le monde n'est donc pas notre lieu naturel. Nous y sommes perdus[106], comme des étrangers en lui. Le monde est par excellence, non notre demeure, mais ce qui ne nous est pas approprié ou familier, l'inhabitable (ἀνοίκειον)[107], ou, dans le meilleur des cas, il n'a qu'un rôle pédagogique[108]. L'âme doit percevoir l'appel venu d'ailleurs qui lui révèle sa vraie nature. La libération apportée par le Sauveur (qui peut être le Christ) libère des puissances du monde, et en particulier de l'influence des astres[109], permettant ainsi la remontée au Père transcendant.

Certains éléments d'une telle attitude connaissent un regain au XIXe siècle avec Schopenhauer, puis avec ses disciples plus ou moins conscients. Une certaine sensibilité de type gnostique renaît. Ainsi, on retrouve l'idée de l'existence comme projection dans le monde chez un personnage de Kierkegaard :

> « Où suis-je ? Que veut dire : le monde ? Que signifie ce mot ? Qui m'a joué le tour de me plonger dans le grand tout et de m'y laisser maintenant ? Qui suis-je ? Comment suis-je entré dans le monde ; pourquoi n'ai-je pas été consulté, pourquoi ne m'a-t-on pas fait connaître les us et coutumes au lieu de m'incorporer dans les rangs, comme si j'avais été acheté par un racoleur de matelots ? »[110].

Ou encore, on trouve l'idée d'un Démiurge méchant ou stupide dans le vers célèbre dans lequel A.E. Housman invite à maudire « whatever brute or blackguard made the world »[111]. Un personnage d'un roman de Joseph Conrad, Stein, associe de façon intéressante deux idées gnostiques, celle de l'étrangeté de l'homme par rapport au monde, et celle de la démence du créateur :

> « Man is amazing, but he is not a masterpiece [...]. Perhaps the artist was a little

---

[104] Basilide, dans Hippolyte, *Réfutations*, VII, 26, p. 374 Duncker.

[105] Clément d'Alexandrie, *Extraits de Théodote*, § 78, 2 ; *Hymne des Naassènes*, v. 7, dans Hippolyte, V, 10 ; cf. aussi *Ginza*, Partie de droite, 11e livre, p. 254 Lidzbarski.

[106] Héracléon, fgt. 23, p. 80 Brooke = Origène, *in Joannem*, XIII, 20.

[107] Clément, *Extraits de Théodote*, § 33, 3, p. 130 Sagnard.

[108] Ptolémée, dans Irénée, *Contre les hérésies*, I, 7, 5 ; SC 264, p. 111.

[109] Clément, *Extraits de Théodote*, § 74, p. 196 Sagnard.

[110] *La reprise*, 2e partie, 11 Octobre, tr. fr. N. Viallaneix, GF, 1990, p. 144.

[111] « The chestnuts are casting their flambeaux », dans *Last Poems* (1922).

> mad. Eh ? What do you think ? Sometimes it seems to me that man is come where he is not wanted, where there is no place for him ; for if not, why should he want all the place ? »[112].

20. L'ALTERNATIVE DES ÉTHIQUES BIBLIQUES

Le monde est créé par un Dieu bon, qui affirme à chaque étape de la création que ce qu'il vient de susciter librement dans l'être est « bon », voire, en son édifice ordonné, « très bon » (*Genèse* 1). Mais les réalités qui semblent les plus sublimes à l'intérieur du monde physique ne sont pas les plus hautes. Elles sont en fait de moindre valeur par rapport à l'homme, qu'elles servent : le soleil et la lune, qui ne sont même pas nommés, sont de simples lampes. L'homme n'a donc pas à se régler sur les réalités du monde, mais doit chercher ailleurs un modèle d'action. Ce modèle est en dernière analyse Dieu lui-même. Celui-ci se manifeste moins par sa création que par une intervention plus directe : il peut soit donner au monde sa loi, comme dans le judaïsme, soit s'y être incarné, comme dans le christianisme. Dans le premier cas, le commandement divin se distingue de tout ce qui est mondain. Ainsi, dans un passage contre le culte des puissances du monde, l'« idolâtrie » :

> « Quant tu lèveras les yeux vers le ciel, quand tu verras le soleil, la lune, les étoiles et toute l'armée des cieux, ne va pas te laisser entraîner à te prosterner devant eux et à les servir. YHWH ton Dieu les a donnés en partage à tous les peuples qui sont sous le ciel. Mais vous, YHWH vous a pris et vous a fait sortir du creuset de fer, l'Égypte, pour que vous deveniez le peuple de son héritage, comme vous l'êtes encore aujourd'hui »[113].

Les corps célestes appartiennent à tous les peuples. Par suite, la Loi ne se trouve pas au ciel[114]. Les commandements imposés à Israël ne dépendent pas du ciel, de la « nature », qui est la même partout et en tout temps, et que chacun peut donc contempler. Ils se fondent sur l'événement unique, historique, de l'exode, qui ne concerne qu'Israël. Le

---

[112] Joseph Conrad, *Lord Jim*, ch. 20, Penguin Classics, p. 195.

[113] *Deutéronome* 4, 19.

[114] *Ibid.*, 30, 12.

Dieu d'Israël n'est pas d'abord celui de la création, mais celui de l'exode[115].

Dans le christianisme, la dévalorisation du monde physique est mise en scène, à travers une imagerie empruntée à la littérature apocalyptique de l'époque, comme une destruction effective, décrite dans le style des prophètes (*Matthieu* 24, 29) ou de la doctrine stoïcienne de la conflagration finale (*2 Pierre* 3, 10, 13). Le monde passe (*4ᵉ Esdras*, 4, 26 ; *1 Corinthiens* 7, 31). La nature, avec ses « signes », n'est pas interprétée comme témoignant de régularités calculables, mais comme annonçant une fin incommensurable (*Matthieu* 16, 2 s.). Il ne s'agit plus de l'imiter, mais d'imiter la bonté de son Créateur (*Matthieu* 5, 45). Le monde est bien plutôt ce à quoi il ne faut pas se conformer (*Romains* 12, 2). La venue du Christ représente l'irruption dans le monde de ce que les puissances qui le gouvernent n'ont pas connu et ne peuvent connaître ; elle établit avec Dieu un nouveau contact qui ne passe pas par le monde. Mourir avec le Christ par le baptême fait mourir aux éléments du monde (*Colossiens* 2, 20), et libère le croyant de l'esclavage qui le leur soumettait (*Galates* 4, 3).

21. EXIGENCES COSMOLOGIQUES DE L'ÉTHIQUE

Quant aux incidences des conceptions du monde sur l'humanité de l'homme, rappelons d'abord une évidence : on ne peut pas démontrer que tout ce dont il a été question ait jamais empêché qui que ce soit de se conduire correctement. Pas plus que, à l'inverse, le modèle pré-moderne du monde aurait jamais détourné personne du vice. En outre, on peut fonder la morale (à supposer qu'elle en ait besoin) sans renvoyer à la question de la structure du monde physique. C'est le cas, par exemple, chez Epicure ou, dans un style très différent, chez Kant. Et Aristote, qui a pourtant marqué plus qu'aucun autre la cosmologie antique et médiévale, ne mentionne à peu près jamais des phénomènes cosmologiques dans son éthique, si ce n'est de façon très allusive et indirecte[116].

La vision moderne du monde a une conséquence importante quant à l'expérience que l'homme fait de sa dimension morale. Tout un aspect de l'homme reste dépourvu de pertinence éthique : sa présence dans le monde. Il n'est pas gênant que l'éthique ne puisse se fonder sur une

[115] Cf. Jehuda Halevi, *Kuzari*, I, § 27, p. 12, 5-7 Baneth.
[116] Cf. *Éthique à Nicomaque*, VI, 7, 1141 a 34-b 2 ; *Métaphysique*, Λ, 10, 1075 a 19 ss.

anthropologie, puisqu'on ne peut de toute façon jamais tirer une norme d'une description. Ce qui gène est que nous ne pouvons plus dire ce que l'éthique a à voir avec le fait que l'homme soit dans le monde. Pourquoi l'homme est-il à la fois éthique et dans le monde ? L'Antiquité n'a guère posé cette question. Pour elle, la réponse étant donnée d'emblée – avec la nature éthique du monde –, la question n'avait pas à être posée. Cette réponse implicite est que la sagesse ne se réalise pas contre la structure du monde, mais est, sinon exigée, du moins encouragée par celle-ci. L'affleurement le plus net de cette question est, à ma connaissance, contenu dans un texte de Plotin, dans son traité contre les gnostiques. Ce n'est pas un hasard, puisque les gnostiques contestaient justement ce présupposé de la pensée grecque. Plotin écrit donc :

> « Si le tout que voici (τόδε τὸ πᾶν) est tel qu'il est permis (ἐξεῖναι), à l'intérieur de celui-ci, et de posséder la sagesse et, tout en étant ici <-bas> de mener une vie (βιοῦν) conforme aux choses de là-haut (κατὰ ἐκεῖνα), comment <ce tout> ne témoigne-t-il pas de ce qu'il est suspendu aux réalités d'au-delà ? »[117].

Plotin partage avec ses adversaires un présupposé : il y a de la sagesse dans le monde. Peut-être choisit-il même, en employant ici le mot *sophia*, rare chez lui, de parler comme les gnostiques, qui racontent les malheurs de Sophie. Celle-ci, comme hypostase divine, est bien présente dans le monde, puisqu'elle y est tombée. La question implicitement posée est transcendantale : que faut-il que soit le monde pour que la sagesse y soit possible ?

Cette question reste fondamentale pour nous aussi. Nous avons vu comment la pensée classique cherchait à mettre en accord la quête morale du bien avec la présence massive de celui-ci dans la nature, conçue et perçue comme un bel ordre, comme un *kosmos*. Nous avons vu comment le pilier physique de ce pont entre l'homme et le monde s'était écroulé. Il nous faut maintenant chercher un autre concept de monde, qui ne sera plus physique, mais qui exprimera la façon même dont l'homme est présent et vit « dans le monde ».

---

[117] *Ennéades*, II, 9 [33], 8, 43-45 ; t. 1, p. 235 Schwyzer-Henry.

22. LE MONDE HUMANISÉ

Une première tentative consiste à constater la présence, à côté du monde naturel ou créé, d'un monde humain. Le mot « monde » signifie aussi, depuis très longue date, la communauté des hommes. Mais cela ne suffit pas pour que la mondanéité de celle-ci soit élevée au concept. C'est ce que fait Vico, qui établit un parallèle strict entre l'histoire humaine et l'univers physique. Il conçoit l'activité historique de l'homme comme constituant un « monde de la cité (*mondo civile*) ». L'homme en est le créateur comme Dieu est celui de l'univers physique. En conséquence, l'homme, qui le crée, et parce qu'il le crée, connaît ce monde comme Dieu connaît le monde de la nature :

> « Ma, in tal densa notte di tenebre ond'è coverta la prima da noi lontanissima antichità, apparisce questo lume eterno, che non tramonta, di questa verità, la quale non si può a patto alcuno chiamar in dubbio : che questo mondo civile egli certamente è stato fatto dagli uomini, onde se ne possono, perché se ne debbono, ritruovare i princìpi dentro le modificazioni della nostra medesima mente umana. Lo che, a chiunque vi rifletta, dee recar maraviglia come tutti i filosofi seriosamente si studiarono di conseguire la scienza di questo mondo naturale, del quale, perché Iddio egli il fece, esso solo ne ha la scienza ; e traccurarono di meditare su questo mondo delle nazioni, o sia mondo civile, del quale, perché l'avevano fatto gli uomini, ne potevano conseguire la scienza gli uomini »[118].

Dans le sillage de Vico, dont il cite l'idée centrale au moins une fois[119], Marx élargit la perspective pour faire de l'homme en son travail le créateur de ce qu'il appelle un monde :

> « Das praktische Erzeugen einer gegenständlichen Welt, die Bearbeitung der unorganischen Natur ist die Bewährung des

[118] *Scienza Nuova* (1744), Livre I, section 3, Des Principes, § 331, p. 479 ed. F. Nicolini, Ricciardi, Milan et Naples.
[119] *Das Kapital*, IV, xv, 1, note.

> Menschen als eines bewußten Gattunsgswesens, d.h. eines Wesens, das sich zu der Gattung als seinem eignen Wesen oder zu sich als Gattungswesen verhält »[120].

Feuerbach n'a pas vu,

> « wie die ihn umgebende sinnliche Welt nicht ein unmittelbar von Ewigkeit her gegebenes, sich stets gleiches Ding ist, sonder das Produkt der Industrie und der Gesellschaftszustandes, und zwar in dem Sinne, daß sie ein geschichtliches Produkt ist, das Resultat der Tätigkeit einer ganzen Reihe von Generationen, deren Jede auf den Schultern der vorhergehenden stand, ihre Industrie und ihren Verkehr weiter ausbildete, ihre soziale Ordnung nach den veränderten Bedürfnissen modifizierte[...] ».

Selon Marx : « diese Tätigkeit, dieses fortwährende sinnliche Arbeiten und Schaffen, diese Produktion <ist> die Grundlage der ganzen sinnlichen Welt ». Il faudrait concevoir « die sinnliche Welt als die gesamte lebendige sinnliche Tätigkeit der sie auszumachenden Individuen »[121].

On remarquera cependant que « monde » désigne alors la nature, ou simplement la terre : ce qui s'appelle production du monde s'appelle aussi ailleurs « Produktion der ganzen Erde »[122]. L'homme dans le monde est « [...] der wirkliche, leibliche, auf der festen wohlgerundeten Erde stehende, alle Naturkräfte aus- und einatmende Mensch [...] »[123]. Marx peut concevoir le rapport de l'homme et de la nature, la sensibilité, le travail, etc., d'une façon originale et féconde ; en revanche, il ne semble pas s'être interrogé sur la mondanité du monde en tant que telle.

23. Le monde intelligible

C'est sans doute Kant qui a refait de la notion de monde l'objet d'une interrogation. Dès la période pré-critique, le monde apparaît comme problématique. La *Dissertation de 1770* voit dans le concept de totalité

---

[120] *Ökonomisch-philosophische Manuskripte* (1844), 1er Ms., xxiv, p. 517.
[121] *Deutsche Ideologie*, p. 41-43.
[122] *Ibid.*, p. 34.
[123] *Ökonomisch-philosophische Manuskripte* (1844), 3e Ms., xxvi, p. 577.

absolue la croix des philosophes[124]. Car cette totalité ne peut jamais être donnée : la marche vers le plus grand ou le plus petit, la régression vers la cause, le parcours de la série des effets ne peuvent être menés à bien. Avec cette constatation, le concept de monde se déplace vers un autre domaine, quittant celui de la contemplation, dans lequel il était installé depuis l'Antiquité : le monde ne peut plus être l'objet d'une expérience, mais seulement celui d'une pensée.

L'œuvre maîtresse procède à la destruction de la cosmologie rationnelle, sous la forme que Wolff lui avait donnée[125], en montrant qu'elle aboutit à des antinomies. Le monde suppose une totalité absolue. Il n'est d'ailleurs pas le seul à l'exiger : toutes les idées transcendantales le font. C'est pourquoi toutes ont droit au nom de « concepts cosmiques » (*Weltbegriffe*)[126] ; et ce que nous appelons « le monde », que l'on peut distinguer aussi plus finement en monde et nature[127], n'est qu'un cas particulier de cette cosmicité plus générale de toute idée. Les idées sont le fait de la raison. Celle-ci ne peut trouver un champ à sa mesure dans le domaine théorique. Elles y sont, avec elle, restreintes à un usage purement régulateur[128].

Mais Kant n'éloigne la raison pure de son escapade théorique que pour la rendre à son domaine authentique, qui est pratique-éthique. Il y retrouve donc, sous un autre mode, ce qu'il avait expulsé du théorique. Les idées de la raison, qui, dans leur usage théorique, étaient confinées dans un rôle régulateur, redeviennent alors législatrices :

> « [...] weil es im praktischen Gebrauche des Verstandes ganz allein um die Ausübung nach Regeln zu tun ist, so kann die Idee der praktischen Vernunft jederzeit wirklich, obzwar nur zum Teil, in concreto gegeben werden, ja sie ist die unentbehrliche Bedingung jedes praktischen Gebrauchs der Vernunft. Ihre Ausübung ist jederzeit begrenzt und mangelhaft, aber unter nicht bestimmbaren Grenzen, also jederzeit unter dem

---

124 *De mundi sensibilis atque intelligibilis forma et principiis*, I, § 2, iii.
125 *Cosmologia rationalis* (1731).
126 Kant, *Kritik der reinen Vernunft*, A 407/B 434.
127 *Ibid.*, A 418 / B 446.
128 *Ibid.*, A 684 / B 712.

> Einfluß des Begriffs einer absoluten Vollständigkeit »[129].

Et c'est justement le cas de l'idée de monde. Le vrai monde est le monde des esprits. Le langage courant, qui parle de « connaissance du monde » alors qu'il n'entend par là que l'expérience des hommes, le pressent[130]. Kant approfondit cette intuition en la fondant sur l'analyse de ce qui fait de l'homme ce qu'il est, à savoir la liberté. Le monde physique est impuissant à être adéquat au concept de monde, il n'est pas assez « mondain ». Kant réalise de la sorte un découplage capital : l'idée de monde est libérée de la physique. Le monde entre dans le domaine éthique, sous la figure du monde intelligible. En soi, l'idée d'un monde intelligible est ancienne. Elle ne se trouve pas chez Platon, qui parle bien d'un « lieu supracéleste », dans le mythe du *Phèdre*[131], mais qui ne fait pas de ce lieu un « monde » ; réciproquement, l'« autre monde » dont parle le christianisme n'est pas le lieu des idées platoniciennes, comme le souligne Origène[132]. Mais l'idée apparaît avec Philon, et, chez Plotin, elle est accompagnée de cette affirmation, capitale pour le lien de l'idée de monde avec la subjectivité, que ce monde intelligible se trouve en nous, et même, que nous *sommes* ce monde[133], idée dont Leibniz avait remarqué l'intérêt[134]. Cependant, si l'idée de « monde intelligible » est ancienne et ambiguë, comme Kant le sait très bien[135], c'est à partir de Kant seul que l'on est à même de penser comme tel le caractère « mondain » de l'édifice des idées.

C'est à partir de l'optique morale que les trois idées que la dialectique transcendantale chassait de l'usage spéculatif de la raison retrouvent leurs droits. Et en particulier celle de monde :

> « Die durch die Achtung fürs moralische Gesetz notwendige Absicht auf das höchste Gut und daraus fliessende Voraussetzung der objektiven Realität desselben führt also durch Postulate der praktischen Vernunft zu Begriffen, welche die spekulative Vernunft zwar als

---

[129] *Ibid.*, A 328 / B 384 s.
[130] *Anthropologie*, Préface, t. VI, p. 399 Weichschedel.
[131] Platon, *Phèdre*, 247 c 3.
[132] Origène, *Des principes*, II, 3, 6, p. 316 Görgemanns-Karpp.
[133] *Ennéades*, III, 4 [15], 3, 22 et IV, 7 [2], 10, 35 s.
[134] *Lettre à Hanschius*, § 3, p. 445 b Erdmann.
[135] *Kritik der reinen Vernunft*, A 256/B 312.

> Aufgaben vortragen, sie aber nicht auflösen konnte. Also [...] führt sie zu dem, wovon die spekulative Vernunft nichts als Antinomie enthielt, deren Auflösung sie nur auf einem problematisch zwar denkbaren, aber seiner objektiven Realität nach für sie nicht erweislichen und bestimmbaren Begriffe gründen konnte, nämlich die kosmologische Idee einer intelligibelen Welt und das Bewußtsein unseres Daseins in derselben [...] »[136].

Le monde physique n'est pas assez solide pour fournir une base permettant l'accès à Dieu, car nous ne pouvons savoir si le monde réel est le meilleur de tous les mondes possibles sans connaître ceux-ci (ce qui est impossible)[137]. En revanche, le monde moral est donné tout d'un coup comme le seul possible. Le vrai monde est le règne des fins. Cela est implicite dans la conclusion de la *Critique de la raison pratique*. La présence de la loi morale en moi :

> « stellt mich in einer Welt dar, die wahre Unendlichkeit hat, aber nur dem Verstande spürbar ist, und mit welcher (dadurch aber auch zugleich mit allen jenen sichtbaren Welten) ich mich nicht wie dort in bloß zufälliger, sondern allgemeiner und notwendiger Verknüpfung erkenne »[138].

Reprenant l'image ancienne de l'amphibie, Kant écrit que nous appartenons à deux mondes[139]. Comment réconcilier ces deux mondes ? Le gouffre entre l'harmonie du monde physique et le désordre de l'histoire humaine semble béant[140]. Kant ne peut qu'ébaucher une réconciliation entre le monde moral et l'univers physique à partir d'une philosophie de l'histoire, et de l'esthétique. La beauté du monde ne pourra servir que comme « symbole de la moralité (*Symbol der Sittlichkeit*) »[141].

La nouvelle détermination du concept de monde résout en tout cas la question de notre présence à celui-ci. On peut bien percevoir dans le

---

[136] *Kritik der praktischen Vernunft*, I, 2, 2, vi, p. 239 s. (pages de l'édition originale)
[137] *Ibid.*, I, 2, 2, vii, p. 251.
[138] *Kritik der praktischen Vernunft*, Conclusion, p. 289.
[139] *Ibid.*, I, 1, 3, p. 155 et *Grundlegung zur Metaphysik der Sitten*, III (AK, IV, 451 s.)
[140] *Ideen zu einer allgemeinen Geschichte in weltbürgerlicher Absicht*, § 9, t. VI, p. 49.
[141] Cf. Kant, *Kritik der Urteilskraft*, § 59.

caractère « contingent » (*zufällig*, « tombé comme cela ») de notre présence dans l'univers physique une trace ultime d'antiques images : celle, gnostique, du jet, ou l'image épicurienne du naufrage[142]. Mais dire que nous sommes dans le monde n'est pleinement vrai que si l'on entend par là le monde intelligible. En effet, si notre présence dans l'univers physique est contingente, notre appartenance au monde moral est nécessaire. En même temps, l'aspect anthropologique du concept de monde prend une nouvelle dimension. En effet, le concept même sur lequel repose l'anthropologie, le concept d'homme, donc, est déterminé à nouveaux frais : la morale ne concerne pas l'homme comme un certain type d'animal, mais tout être rationnel en général[143], dans la mesure où, s'il est fini, un être de ce genre éprouvera la loi comme quelque chose d'extérieur.

24. LE CONCEPT PHÉNOMÉNOLOGIQUE DE MONDE

Un approfondissement analogue du concept de monde est proposé, en un autre style, par Heidegger. Celui-ci ne part pas du rapport perceptif avec les choses. Pour lui, ce rapport n'est pas premier, il est sous-tendu par un rapport d'usage avec des ustensiles ; le monde est bien alors l'objet d'une expérience, mais cette expérience n'est pas perceptive, elle est celle d'une habitation. Elle est celle d'une présence totale, d'un tout ou rien. Le concept de monde devient ainsi subjectif, ou plutôt il est constitutif de la subjectivité.

L'analyse de l'idée de monde par Heidegger occupe une période qui s'étend environ sur dix ans. Il commence par l'analyse de l'idée de « monde environnant » (*Umwelt*), de ce qui – Heidegger forge un verbe pour le dire – « monde » (*es weltet*)[144]. Il en dégage l'idée selon laquelle la vie est essentiellement marquée par le fait de vivre dans le monde (*Weltcharakter des Lebens*) :

> « Unser Leben ist die Welt, in der wir leben, in die hinein und je innerhalb welcher die Lebenstendenzen laufen. Und unser Leben ist nur als Leben, insofern es in einer Welt lebt. [...] Unser faktisches Leben ist unsere Welt [...]. Die

---

[142] Cf. supra n. 105 et Lucrèce, V, 222 s.
[143] Kant, *Kritik der praktischen Vernunft*, § 7, p. 57 et cf. *Kritik der reinen Vernunft*, B 72.
[144] *Zur Bestimmung der Philosophie*, cours SS 1919 (GA 56/57), p. 73.

> faktische Lebenserfahrung ist im wörtlichen Sinne " weltlich gestimmt ", sie lebt immer in eine Welt hinein, sie befindet sich in einer Lebenswelt »[145]. « Welt ist die Grundkategorie des Gehaltsinnlichen im Phänomen Leben »[146].

L'œuvre maîtresse de 1927, *Sein und Zeit*[147], fournit une première synthèse avec son analyse de l'« être-dans-le-monde (*In-der-Welt-sein*) ». Celui-ci n'est pas une catégorie, une caractéristique de ce qui est disponible (*vorhanden*), mais un existential, une caractéristique du *Dasein*. Il faut distinguer entre le monde et la nature. Toute l'ontologie passée a enjambé le phénomène du monde : elle confondait celui-ci avec la nature, qu'elle prenait pour fil conducteur[148]. Or, la nature n'est pas le monde, mais un étant qui se rencontre à l'intérieur du monde. En revanche :

> « " Welt " ist ontologisch keine Bestimmung *des* Seienden, das wesenhaft das Dasein *nicht* ist, sondern ein Charakter des Daseins selbst »[149].

Vom Wesen des Grundes (1929) approfondit l'analyse, en fournissant en outre quelques éléments sur l'histoire du concept :

> « Das Dasein ist nicht deshalb ein In-der-Welt-sein, weil und nur weil es faktisch existiert, sondern umgekehrt, es kann nur als existierendes sein, d.h. als Dasein, weil seine Wesensverfassung im In-der-Welt-sein liegt » [...]. « Es is daher gleich irrig, den Ausdruck Welt entweder als Bezeichnung der Allheit der Naturdinge (naturaler Weltbegriff) oder als Titel für die Gemeinschaft der Menschen (personaler Weltbegriff) in Anspruch zu nehmen. Vielmehr liegt das metaphysisch Wesentliche der mehr oder minder klar abgehobenen Bedeutung von κόσμος,

---

145 *Grundprobleme der Phänomenologie*, WS 1919-1920 (GA 58), p. 33 s., 96, 250.

146 *Phänomenologische Interpretationen zu Aristoteles*, cours WS 1921 - 1922 (GA 61), p. 86.

147 L'œuvre est préparée par tous les cours qui précèdent, mais plus particulièrement par *Ontologie*, cours SS 1923 (GA 63), §§ 16-26, p. 79-104 et *Prolegomena zur Geschichte des Zeitbegriffs*, cours SS 1925 (GA 20), §§ 18-26.

148 Cf. mon « La phénoménologie... ».

149 *Sein und Zeit*, § 14, p. 63 s.

> *mundus*, Welt darin, daß sie auf die Auslegung des menschlichen Daseins *in seinem Bezug zum Seienden im Ganzen* abzielt » [...]. « Zur Selbstheit gehört Welt ; diese ist wesenhaft daseinsbezogen [...] »[150].

C'est peut-être dans le cours du semestre d'hiver 1929-1930 que Heidegger pousse le plus loin son analyse du phénomène du monde, qu'il relie explicitement aux réflexions développées dans *Vom Wesen des Grundes*. Il le fait à partir d'une comparaison entre minéral, animal et humain qui développe une triple thèse : « 1. der Stein (das Materielle) ist weltlos ; 2. das Tier ist weltarm ; 3. der Mensch ist weltbildend »[151]. Le dernier point n'a pas été l'objet d'un développement aussi minutieux que le second, et reste, me semble-t-il, programmatique. La pensée de Heidegger ne reviendra ensuite vers la question du monde qu'indirectement, et à un autre niveau.

Je me contenterai donc de remarquer que, dans cette optique également, et d'une façon au moins aussi radicale que chez Kant, le concept d'homme est repensé et enlevé à toute « anthropologie ». Et c'est la présence dans le monde qui permet de repenser ainsi l'humanité de l'homme. Une fois pensé comme *Dasein*, comme être-là, le monde est conçu à partir du « là » de l'homme. Mais être-« là » est l'objet d'une invitation adressée à l'homme, du fait « daß dem Menschen das Dasein als solches zugemutet wird, daß ihm aufgegeben ist – da zu sein » [...]. « Der Mensch, wenn er werden soll, was er ist, je gerade das Dasein sich auf die Schulter zu werfen hat [...] ». Il s'agit de « die Menscheit im Menschen befreien, die Menschheit des Menschen, d.h. das Wesen des Menschen befreien, das Dasein in ihm wesentlich werden lassen. Diese Befreiung des Daseins im Menschen heißt nicht, ihn in eine Willkür stellen, sondern dem Menschen das Dasein als seine eigenste Bürde aufladen »[152].

25. Le monde subjectif

Ainsi, le concept de monde vers lequel s'oriente la philosophie contemporaine est, en un sens, tout trouvé. Il n'a rien d'original,

---

[150] Repris dans *Holzwege*, 35-55, cité : p. 37, 51 s., 53. Première version dans *Anfangsgründe der Logik*, cours SS 1928 (GA 26), § 11, p. 203-252.

[151] *Grundbegriffe der Metaphysik*, cours WS 1929- 1930 (GA 29/30), p. 261 s., 263.

[152] *Ibid.*, p. 246-248 et cf. p. 255.

puisqu'il correspond à ce que nous visons plus ou moins explicitement quand nous disons, sans trop y réfléchir, que l'enfant qui naît « vient au monde », ou que mourir, c'est quitter le monde. Ces expressions ne se rencontrent d'ailleurs qu'à partir d'une époque déterminée de l'histoire de la pensée, et tout spécialement lors de l'entrée des représentations juives et chrétiennes dans la culture d'expression grecque. Heidegger remarquait que c'est avec le christianisme que : « Die Selbstwelt als solche tritt ins Leben und wird als solche gelebt »[153]. L'expression « entrer dans le monde » se trouve en effet dans le célèbre prologue du quatrième Évangile, où il est question, selon les interprétations, de « tout homme » ou du Verbe divin « qui vient dans le monde » (*Jean* 1, 9). Certaines langues empruntent à cette expérience leur mot pour « monde », en le dérivant de l'idée de durée de vie (grec αἰών), d'« âge d'homme » : ainsi, les langues germaniques qui forment *world*, *Welt*, *vereld* sur les racines pour « homme » (latin *vir*) et pour « âge » (cf. anglais *old*, allemand *alt*) ou encore, le chinois *shì* (japonais *se*).

Essayons pour finir d'élaborer, de façon très élémentaire, les implications de ce concept de monde. Nous entendons aujourd'hui par « venir au monde » ou « quitter le monde », la plupart du temps, entrer dans la compagnie des hommes, ou en sortir. Mais nous percevons toujours que notre venue au monde nous mène à une totalité, et qu'à l'inverse c'est d'une totalité que la mort nous fait prendre congé. Et l'on trouve des penseurs qui élèvent ce sentiment à une formulation explicite : ainsi Montaigne : « Comme nostre naissance nous apporte la naissance de toutes choses, aussi fera la mort de toutes choses, nostre mort »[154]. Ou encore Pascal : « Chacun est un tout pour soi-même, car, une fois mort, le tout sera mort pour soi »[155]. Pascal se replace ici, comme souvent, dans le sillage de Montaigne, même s'il n'envisage que la mortalité et passe sous silence la « natalité » (H. Arendt) de l'homme. Mais sa formule, outre une plus grande prégnance stylistique, a le mérite de faire passer l'idée de totalité du côté du sujet. Montaigne parle de « toutes choses » ; Pascal parle du tout que nous sommes, et qui fait que les choses sont pour nous une totalité. Le caractère total des choses, qui fait de celles-ci le monde qu'elles forment, tient à notre mortalité, ou à son anticipation.

---

[153] Heidegger, *Grundprobleme der Phänomenologie*, WS 1919-1920 (GA 58), p. 61.
[154] *Essais*, I, 20 ; t. 1, p. 166 Villey (1930).
[155] *Pensées*, Br. 457/Lafuma 668.

Il faut du tout pour faire un monde. Mais la totalité dont il est ici question n'est pas obtenue par une synthèse. Elle est donnée d'emblée. Pourtant, sa totalité n'est pas quantitative. L'unité ne résulte pas du parcours de la multiplicité, comme son épuisement. Elle précède et rend possible tout parcours, et subsiste en tout point de celui-ci. Le caractère total de la présence au/du monde n'a de la sorte rien à voir avec la quantité d'objets dont nous faisons l'expérience, qui est bien évidemment une partie infime de tout ce que contient le monde. De la sorte, le monde n'est pas monde par soi, sa « mondanité » ne vient pas de lui, mais de nous.

S'il en est ainsi, mettre l'idée de monde en rapport avec l'anthropologie ne consiste nullement à rejeter l'infini de l'univers au profit de la chaude intimité de l'environnement. En entrant dans le monde, à notre naissance, nous entrons dans ce qui contient tout aussi décidément, mais aussi indifféremment, notre berceau que les plus lointaines nébuleuses. Plus radicalement, il ne s'agit pas non plus de se réfugier dans la subjectivité en abandonnant le monde des objets « concrets ». C'est l'inverse : il s'agit de montrer que le vrai monde est situé du côté du sujet, et que ce que nous appelons « le monde » – soit l'univers des objets – n'est pas capable de satisfaire aux réquisits du concept de monde, en un mot : n'est pas encore assez « mondain » pour l'homme.

Ainsi, nous avons vu d'abord comment l'idée de monde, dès sa formulation grecque comme *kosmos*, impliquait une anthropologie déterminée. Nous avons vu ensuite comment ce qui restait virtuel s'est développé, dans l'Antiquité et au Moyen Age, en une anthropologie cosmologique ou une cosmologie anthropologique. Puis, nous avons vu comment ce lien s'est défait à l'époque moderne. Nous avons enfin vu comment il pourrait se reformer, voire se resserrer encore plus étroitement. Mais ce serait au prix d'une double reformulation de l'idée de monde et de celle d'homme, pensées dans leur appartenance réciproque. Une telle pensée permettrait de concevoir dans quelle mesure il n'est d'homme que du monde.

BIBLIOGRAPHIE

ALLERS Rudolf, « Microcosmus from Anaximander to Paracelsus », *Traditio*, 2 (1944), 319-407.

ASSMANN Jan, *Ma'at*. Gerechtigkeit und Unsterblichkeit im alten Ägypten, Beck, Munich, 319 p.

BLUMENBERG Hans, *Die Genesis der Kopernikanischen Welt*, 3 vol. (stv n° 352), Francfort, 1981, 803 p.

BRAGUE Rémi, « La phénoménologie comme voie d'accès au monde grec. Note sur la critique de la *Vorhandenheit* comme modèle ontologique dans la lecture heideggerienne d'Aristote », J.-L. Marion/G. Planty-Bonjour (edd.), *Phénoménologie et Métaphysique*, P.U.F., Paris, 1984, 247-273.

– « Vers un concept de l'être-au-monde : la presse (θλῖψις) », P.A. Simon (ed.), *Pela Filosofia*. Homenagem ao Prof. T.M. Padilha, Pallas, Rio de Janeiro, 1984, 229-241.

– *Aristote et la question du monde*. Essai sur le contexte cosmologique et anthropologique de l'ontologie, P.U.F., Paris, 1988, 560 p.

– « L'expérience biblique du monde et l'idée de création », *Création et salut*, Publications des Facultés Universitaires Saint-Louis (n° 47), Bruxelles, 1989, 105-120.

- « Geozentrismus als Demütigung des Menschen », *Internationale Zeitschrift für Philosophie*, 1/1994, 2-25.

– « Cosmos et éthique. La fin d'un modèle », *Acta Institutionis Philosophiae et Estheticae* (Tokyo), vol. 12, Eco-ethica et civilisatio moderna (1994), 53-64.

– « Deux versions du microcosme. Etre le monde en petit ou devenir le monde en grand », A. Hasnawi et al. (edd.), *Perspectives arabes et médiévales sur la tradition scientifique et philosophique grecque* [...], Peeters, Louvain, 1997, 523-533.

– « Une mystique cosmologique. L'imitation des corps célestes dans le Hayy ibn Yaqzân d'Ibn Tufayl », P. Fenton et R. Goetschel (ed.), *Écriture et expérience mystique dans les trois religions du livre*, Leyde, Brill, 2000, 89-100.

CHIESA Bruno, *Creazione e caduta dell'uomo nell' esegesi giudeo-araba medievale*, Brescia, 1989, 209 p.

FESTUGIÈRE André-Jean, o.p., *La révélation d'Hermès Trismégiste*, II. Le Dieu cosmique, Gabalda, Paris, 1949, XVII-610 p.

GRANT Edward, « Medieval and Renaissance Scholastic Conceptions of the Influence of the Celestial Region on the Terrestrial », *Journal of Medieval and Renaissance Studies*, 17 (1987), 1-23.

HALLYN Fernand, *La structure poétique du monde : Copernic, Képler*, Seuil, Paris, 1987, 316 p.

KOYRÉ Alexandre, *From the Closed World to the Infinite Universe*, Johns Hopkins Press, Baltimore, 1957, XII-313 p.

LEWIS Clive Staples, *Studies on Words*, Cambridge U.P., 1967 (2e ed.), VII-343 [ch. 9 : WORLD, 214-268].

LLOYD Geoffrey E. R. « Greek Cosmologies », *Methods and Problems in Greek Science*. Selected Papers, Cambridge U.P., 1991, 141-163.

LÖWITH Karl, « Mensch und Menschenwelt » (1960), dans *Mensch und Menschenwelt*. Beiträge zur Anthropologie (Sämtliche Schriften, I), Metzler, Stuttgart, s.d., 295-328.

MERLEAU-PONTY Jacques & MORANDO Bruno, *Les trois étapes de la cosmologie*, Paris, 1971, 319 p.

NORTH J. D., « Medieval Concepts of Celestial Influence : A Survey », Patrick Curry (ed.), *Astrology, Science and Society*. Historical Essays, Boydell Press, Woodbridge / Wolfeboro, 1987, 5-17.

PELLEGRINO Michele, « Il " topos " dello " status rectus " nel contesto filosofico e biblico (A proposito di Ad Diognetum 10, 1-2) », A. Stuiber & A. Hermann, *MULLUS*, Festschrift Theodor Klauser, JAC, Ergänzungsband, 1 (1964), 273-281.

RICO Francisco, *El pequeño mundo del hombre*. Varia fortuna de una idea en las letras españolas, Edición corregida y aumentada, Alianza Editorial, Madrid, 1988, 380 p.

SAMBURSKY Shmuel, *The Physical World of Late Antiquity*, Londres, 1962, XII-189 p.

SLOTERDIJK Peter, *Weltfremdheit*, Suhrkamp, Francfort a. M., 1993, 381 p.

NOTE

Commandé en 1993, ce texte constituait une première ébauche d'un travail alors en cours. Celui-ci a été publié sous le titre *La Sagesse du monde. Histoire de l'expérience humaine de l'univers*, Paris, Fayard (« L'esprit de la cité »), 1999. Une version du même livre en format de poche, légèrement augmentée, a été publiée en 2002 à Paris, Hachette (« Biblio-essais »). L'ouvrage a été traduit en portugais à Lisbonne, Piaget, 2003 et en anglais à Chicago, The University of Chicago Press, 2003. Il est en cours de traduction en allemand à Munich, Beck. Le lecteur voudra bien s'y reporter pour y trouver une version développée de ce que je n'avais pu ici qu'esquisser.

PHILOSOPHY OF TECHNOLOGY

Don IHDE
State University of New York
Stony Brook, New York

SUMMARY

I have elsewhere argued that contemporary philosophy of technology has arisen and grown out of the ' praxis ' traditions, particularly those of a concretist orientation, and thus stand in contrast to the earlier, dominant strands of a theoretically biased philosophy of science[1]. And, even if much contemporary philosophy of science has been late to arrive at such praxis phenomena as experiment, instrumentation and technologization, in science, it, too, has begun to take a similar direction. This has some implication for the role of the philosopher of technology or of *technoscience* as current coin would have it.

First, there is some degree to which the philosopher of technology must « go native », by this I mean become more than a distant observer, to become an informed *participant*. Without this participant-observation, the philosopher could never deal with the developmental phases of technologies, which I have argued are as, if not more, important than the response phases which deal with already extant technologies and their effects.

Second, a praxis orientation is necessarily more ' pragmatic ' and area or regionally focused than a high altitude and general theory might be. I see nothing wrong with focused specialization directed towards the various areas of the technologies of the times.

Third, as indicated above, a classical role for philosophers of technology remains conceptual in the sense of re-conceiving or re-

[1] Cf. Don IHDE, *Technics and Praxis*, and *Technology and the Lifeworld*, Indiana University Press, 1990.

P. Kemp (ed.), World and Worldhood, 91-108.

describing phenomena. In this sense one positive feature arising from postmodern sensibility is the appreciation for alternative frameworks and the « fusing of horizons » in a Gadamerian fashion.

Finally, philosophy of technology is necessarily concretist or ' materially ' oriented insofar as the technologies operate materially at whatever level. Such material operations display patterned, structured, and while multistable, *limited* sets of possibilities. It is this structure that philosophers may examine and analyse.

All of this characterizes a certain style of philosophical approach which is beginning to show itself in the new sub-field of the philosophies of technology.

« Philosophies of ... », are historically Hegelian originated. By this I mean that thematic studies which foreground some sub-division of philosophy, usually disciplinarily focused, begin to appear with G.W.F. Hegel (1770-1831), such as may be found in his monographs on *Geschichtesphilosophie*, or his *Religionsphilosophie*. (He did not develop a *Techniksphilosophie*). Indeed, the first explicit title which employs the term « philosophy of technology » comes from the 1877 book published by Ernst Kapp (1808-1896), *Grundlinien einer Philosophie der Technik*. Kapp was one of the 19th century « left Hegelians », as was his slightly younger peer, Karl Marx (1818-1883), who similarly must be noted to stand at the origins of what was to become *philosophy of technology* as a thematic, sub-disciplinary field within Philosophy.

This does not mean that attention to ' technologies ' are entirely lacking in the previous history of philosophy. Those who prefer « master narrative » histories of philosophy – including Martin Heidegger (1889-1976) who is most frequently cited as the twentieth century progenator of the philosophy of technology – tend to include the Classical Greek concern with *techne* as part of this history. The material artifacts, tools, and other ' produced material objects ' have often enough played a background role in philosophy.

Interestingly, however, the period in ancient times which most approximates what later is identified as the birth of Modern Science (in the seventeenth century) remains mostly neglected by even the master narrative philosophers – and that is the ' technoscience ' period of later antiquity, the Hellenic or Alexandrian period of the second and first centuries BCE. That neglected era was, in fact, the most prolific with respect to both the anticipations of Modern Science, if by that is meant the development of actually measured and experimentally verifiable theory, and of Technoscience, which means to be a science embodied in instruments and to develop engineering technologies. Much of this movement comes out of neo-Aristotelianism and is associated with the expansionist period of Alexander the Great[2].

I shall not, however, follow a master narrative development here, but accept the convention that it is with the birth of Modern Science and in its wake, that late modern « philosophies of ... » begin to include, first, the *philosophy of science*, and then later, *philosophy of technology*. The early Modern period associated with the origins of Modern Science may be seen to be, in practice, a science which is *technologically embodied.* Renaissance science is both thoroughly engineering oriented and a ' theory ' which recognizes this is promulgated by Francis Bacon (1561-1626). Bacon not only conceived science to be world-transformative – « knowledge is power », – but explicitly linked this to instrumentation or technologies. He developed within his own system some twenty-seven uses for instrumentation[3]. Still, technologies remain more implicit than explicit. And this also remains the case within the mainstream philosophy. So, again, I return to the post-Hegelian development. Within this perspective, the anticipatory moments which become the philosophy of technology, are, indeed, neo-Hegelian.

If Kapp remains the first to use the title, one cannot ignore the still earlier work of Karl Marx which does begin to make the movements which forefront technological developments, and even in one sense invent « technological determinism ». Both Kapp and Marx were ' materialist ' Hegelians and thus technologies become much more thematic in both their developments.

As early as *Das Kapital* (1867) a kind of technological determinism begins to make entry into the Marxian interpretation of dialectical materialism. What differentiates the different periods in Marxian history, are *concretely developed, fully technologized* « modes of production ». Differences between feudal, capitalistic, and socialistic (communism) periods, are associated with and produced by different systems and arrangements of technologies within differing productive modes. Different tools and tool systems define feudal as contrasted to

---

[2] Cf. L. SPRAGUE DE CAMP, *The Ancient Engineers*, New York, Dorset Press, 1963.

[3] Cf. « Baconian Topics », in Ian HACKING, *Representing and Intervening*, Cambridge University Press, 1983.

industrial productive systems. The social-political results which create « alienation », « class struggle », « de-skilling », and other well known Marxian concepts, all arise within the context of a political-economic-*technological* history. These themes continued to mark much of twentieth century explicit philosophy of technology. In the mid to late twentieth century these themes are associated with the strands of Critical Theory within the « Frankfort School ».

Kapp's approach often parallels Marx's and also concretizes and forefronts the roles of technologies in human history. But if Marx's dominant trajectories are econo-political, Kapp's trajectory is more geographical-territorial. Kapp took « colonization » as a major metaphor and argued that there is both an « internal » and an « external » colonization which characterized the human development of technologies. Technologies were analogized as extensions and magnifications of human organic processes and projected into an external environment. While the analogies to muscle power for various tools clearly extend specific human powers, Kapp also analogized technologies into vaster and more complex systems – for example, communications systems are analogized upon the nervous system, etc. In some sense, then, technologies are material extensions of human embodiment. Kapp, too, was to have echoes in the later twentieth century, most notably in the traditions associated with Marshall McLuhan and James Feibleman.

These paired neo-Hegelian ' materialist ' movements took their place in the 19th century era of the Industrial Revolution. Marx's more negative take upon industrial capitalism, and Kapp's somewhat more positive take upon technological expansionism, are but two sides to a similar development of early technological determinisms.

At this juncture I wish to introduce a sub-set of themes which also characterizes much of the history of the philosophies of technology. As technologies become more forefronted within philosophical reflection, it is clear that there can be both a positive and a negative characterization of the same ever more powerful and explicit phenomenon. Extrapolated, « utopian » directions find counterpart « dystopian » directions. One may find these even earlier than the explicit neo-Hegelian « philosophies of... ». An excellent example of the tension between utopian and dystopian interpretations may be found in the very heart of early Modernity. Francis Bacon, who clearly took a more utopian direction in taking science as means of relieving human suffering through manipulation of the natural world, finds his dystopian counter-part in his Cantabridgian contemporary, Christopher Marlowe (1564-1593), author of *Doctor Faustus*. Here the very attempt to technologically manipulate the natural world is taken as a pact with the Devil or a form of hubris which eventually condemns the user.

These themes, too, continue to find echo in more contemporary work. The Faustus theme becomes the Frankenstein theme which sees in technology a potential run-away autonomy. But the association is one which sometimes juxtaposes what I shall call ' technoscience proponents ' with ' humanist detractors '. This association is echoed in the most thorough history of the philosophy to date, Carl Mitcham's *Thinking Through Technology: The Path Between Engineering And Philosophy* (Chicago, 1994). Mitcham distinguishes between two traditions in philosophy of technology, the ' engineering ' tradition, which he sees as more positive towards technology, and the ' humanities ' tradition which is more critical.

If the appearance of a ' history ' of the philosophy of technology is clear evidence of the contemporary existence of a relatively new « philosophy of ... », along with a small list of various introductions to philosophy of technology[4], this making of ' histories ' and of ' introductions ' is strictly a twentieth century phenomenon. This is understandable precisely because two of the macro-events of modern times relate to the technologization of the work process in the Industrial Revolution, and the even more tragic technologization of war in World War I. Philosophers could hardly ignore such massive transformations of society.

The inter-war period (1919-1938) did see another impetus to the philosophy of technology, both in Europe and in America. If Mitcham is correct regarding two traditions in philosophies of technology – engineering and humanities – it is this formative period which sees this emergence. In 1927 a second book titled, *Philosophie der Technik*, was published by the engineer-philosopher, Friedrich Dessauer (1881-1963). Dessauer was by profession an ' applied scientist ' who later specialized in the application of x-rays to medical uses (and who, himself, died of cancer induced by self-experimentation). Philosophically, Dessauer was

---

[4] Introductions to the field include : Frederick FERRE, *Philosophy of Technology*, Prentice Hall, 1988 ; Jean-Yves GOFFI, *La Philosophie de la technique*, Presses Universitaires de France, 1988 ; Don IHDE, *Philosophy of Technology : An Introduction*, Paragon House Publishers, 1993.

a neo-Kantian and his analyses often took the Kantian distinction between phenomena and *noumena* to be of importance. But – and this was yet another move in making the technological processes forefront processes – Dessauer argued that if epistemology must be limited to phenomena (appearances), the making and using of technologies or material production gets to the *noumena.* And much of his analysis then turned to classifications of technologies and technological processes.

Interestingly, 1927 is also the year of the publication of *Being And Time* by Martin Heidegger. One cannot claim *Being And Time* is a ' philosophy of technology ', but it does contain what can be seen in retrospect are *proto-technological* themes. This is particularly the case with respect to the relationship between *Vorhandenheit* and *Zuhandenheit*, or between pragmatic praxis in and through tools and the dependent or secondary role by which scientific or ' objective ' objects are constituted[5].

In North America John Dewey (1859-1952) was similarly engaged in developing a primacy of praxis orientation to philosophy itself. His « instrumentalism » or pragmatism, he said later in life, he should have called « technology »[6]. Dewey modelled philosophy itself upon the pragmatic, problem solving process. He was already a ' non-foundationalist ' with regard to the previous histories of philosophy, and saw the need to constantly adapt critical thinking to its contemporary context and problems. The themes were quite explicit as early as *The Quest For Certainty* (1929).

In the thirties, two other thinkers began to take very explicit account of the role of technology in human affairs: Jose Ortega y Gasset (1883-1955) and Lewis Mumford (1885-1988). In his 1933 lectures, published in 1935, Ortega argued for an interactive concept of the self. « I am I plus my circumstances ». And although the emphasis upon technologies as primary components of « my circumstances » grew in later life, they were included from the beginning in this contextual mode of human interpretation. Mumford, with *Civilization And Technics* (1934), argued for the central role technologies and their development played in human history. His analysis of the introduction of the

---

[5] Cf. « Heidegger's Philosophy of Technology », in Don IHDE, *Technics and Praxis : A Philosophy of Technology*, Reidel, 1979.

[6] Cf. Larry A. HICKMAN, *John Dewey's Pragmatic Technology*, Indiana University Press, 1990.

mechanical clock into Western practice remains a paradigm issue. *Homo faber* was concrete humanity and develops in relation to its technologies.

Although there are other figures who might be mentioned in the inter-war period, the strand of philosophy which arises from and within a ‘ technical ’ context and represented above by Dessauer, does begin to have a counter-part. Heidegger, Ortega, and Mumford begin a non-technical, perhaps better termed a non-technocratic, tradition of ‘ humanities ’ perspectives upon technology.

All this takes shape against the rising impact of larger and more complex technologies which are increasingly being incorporated into the Western practice. The late nineteenth century was marked by the Industrial Revolution as noted and here Marx emerged as its most eminent respondent. The twentieth century saw another macro-technological revolution in its world wars.

In the broadest sense, there have always been military technologies. David's slingshot versus Goliath's spear is but an ancient biblical example. But World War I was a dramatic event which so well illustrates the issues of technological transformation, and transformation on a grand scale, in which deeply entrenched practices were seen to contrast with new technologies such that the historical force of technologies could not be ignored. Moreover, the contrast between the outmoded strategies still favored by traditionally trained military commanders and the *ad hoc* strategies which had to be invented for the new technologies produced so much of the death and senseless suffering of that War. The invention of the airplane only eleven years prior to the War, the machine gun and the tank, all injected into battlefields still governed by infantry and cavalry charges and massed formations, simply provided more human ‘ material ’ for the new technologies. Yet, World War I had not yet reached a synthesis between social praxis and technologicial complexes. World War II did – it was the ‘ Industrial ’ or factory War. If its most tragic form was the mass means of slaughter as in the Holocaust, its engines were the Industrialized powers which drove nations to produce virtually infinitely expendable machines (+ humans) to feed their war efforts.

It is not surprising, then, that in the wake of World War II there arose a multitude of philosophers with critiques of a new technologized civilization. In the post-World War II period, from the forties through the sixties, the dominant European themes ran from the distinctly dystopian through the recognition that there must be some corrective ' fix ' applied to technologization.

Post-World War II philosophies which responded to technology in Europe might be grouped in three very broadly defined traditions: (1) The Marxian tradition was re-formulated within ' Critical Theory ', usually associated with the Frankfort School and whose main proponents were Max Horkheimer, Theodor Adorno, Jürgen Habermas and Herbert Marcuse. (2) The ' Existential ' traditions which often overlapped in shared critiques with the Critical Theorists, could be thought to include Martin Heidegger, Karl Jaspers, Gabriel Marcel, Ortega y Gasset and even Nicolas Berdyaev. And, if conceived broadly enough, one should also consider the sociologist-theologian, Jacques Ellul, to belong here. Both these broad traditions were, in Mitcham's classifications, ' humanities ' philosophies and tended to be critical. (3) The ' engineering ' traditions are, perhaps, a bit less known, but include such figures as Gilbert Simondon and a group of philosophers associated with the post-war *Verein Deutscher Ingenieure* (*VDI*). This latter tradition clearly recognized human problems arising out of technologization, but tended to take an « ethics » or ' repair ' approach to technology.

I shall not, however, trace the post-war philosophy of technology developments by traditions alone. Instead, I wish to focus upon what could be called common themes and problems which often overlap traditions. Each of these problems received very wide attention and discussion from the mid-forties on through the sixties. But in each case, whether conservative in the sense of looking for correctives, or radical in the sense of seeing in Technology a run-away, autonomous phenomenon, each response is one which sees technologization having produced or having helped produce historically negative effects.

A ' CONSERVATIVE ' CORRECTIVE APPROACH

I begin with what might be called the most ' conservative ' response to the negative effects of technologization as part of a response to the mass destructiveness of World War II. In contrast to the often utopian hopes of the 19th century which saw in technologization the hopes for humanity's social problems' solutions, the disaffection which followed the Wars called into question the naive and technocratic directions of the pre-war environment. The ' conservative ' response was one which implicitly often takes the following approach to technologization :

(1) Technology or technologies may remain thought of as « neutral ». Technologies are simply human inventions which get used in good or bad ways. Admittedly, larger and more powerful technologies and more complex systems of technologies are potentially either more likely to have larger effects, whether good or bad. Thus, the negative effects of technologization in the Wars were to be blamed on the human – that is the ethical or unethical – uses which the makers and/or the politically powerful could make of them.

(2) What may have made the War effects worse, however, was the extreme technicization of the professions and disciplines themselves. Technical education, industrial organization, and even the concepts of ' applied ' sciences were often technocratic and bereft of taking into account the ' human ' side of engineering or the social-political effects of the same.

(3) Thus the obvious corrective is also seen to be a kind of ' ethics fix '. What is needed is to critically reflect upon technologies, their likely effects, and to introduce into the situation the deeper thinking that is called forth to humanize or ethicize the inventors, producers, users and controllers of these technologies. It was in this context that the *Verein Deutscher Ingenieure*, in particular, held conferences beginning in 1947 and continuing in various respects to the present. It is also in this context that the concept and development of various « technology assessment » analytic tools were thought to be helpful correctives.

(4) Such responses, however, also have implications for the role of the philosopher and for philosophy of technology. In the most conservative sense, this means that philosophers of technology are *reactive*. Technologies are left in the domain of the inventors, developers, and then the users and controllers, but technological development is implicitly left to the ' technologists '.

One late twentieth century institutionalization of this approach has been the development of institutionalized « ethics committees », particularly in Medical Schools and Hospitals. ' Ethicists ', a late twentieth century term, or ethicists in committees of experts, laymen plus ethicists, respond to crises, shortage, and other situations which arise from the development of new technologies. In America one of the earliest such developments occured in response to early kidney dialysis technologies in the sixties and has grown into a kind of sub-discipline which extends beyond the medical into various professional contexts and is usually termed « applied ethics ».

(5) One other characteristic of the ' conservative ' or corrective approach is that there is a tendency for the philosophers involved – although this is not a necessary response – is for them to themselves adapt to a quasi-technocratic or analytic approach. Ethics committees often develop various means by which ' calculii ' of considerations, such as those often associated with the various brands of *utilitarianism* may be used. Not unlike technology assessment, ethics committees can utilize such means and thus become simulacra of the institutions within which they work.

(6) One distinct advantage of the corrective approach is that it necessarily has to pay attention to particular technologies or technology complexes. Thus, if the situation is one of cost and scarcity (as in early kidney dialysis), the outcomes are specific to the complex involved and the evaluative personnel gradually build up an expertise regarding technologies and their effects.

A SOCIAL-CRITICAL APPROACH

A centrist approach to technologization is one which concentrates upon social, cultural, and political outcomes. In this case what I am calling the social-critical approach is one whose elements may be found in several traditions :

(1) In its most negative sense, a social-critical response to technologization may be detected both in the early Critical Theorists (Adorno and Horkheimer) and amongst some of the ' existentialist ' oriented thinkers (Berdyaev, Heidegger, and Ortega). Here the claim was that technologization posed a threat to individuation and to aspects of creativity thought to belong to individuation. Technologized culture was thought to be ' mass culture ', and to lead to a kind of cultural levelling. Heidegger's *das Mann*, Berdyaev's loss of *sobornost* (community), Ortega's ' mass man ', are variants upon this theme.

(2) Not only do the social-critical thinkers hold that technologization leads to a levelling and anonymous public phenomenon, but this levelling echos the earlier Marxian claims that industrialization de-skills producers and thus alienates them from their work. *The often romantic theme regarding craft degradation in various forms frequently characterizes technology criticism.* Heidegger's romanticism over hand-

work tools, including his criticism of the typewriter compared to the pen exemplifies this tendency[7].

(3) If craft work de-skilling is laid to an industrially embedded technologization, the same social-criticism applies to the ‘ high arts ’ of European cultural values. Ironically, it is the neo-Marxian Critical Theorists – most particularly Adorno and Horkheimer – who decry this presumed effect of technogization. ‘ Popular culture ’, is taken to be a degradation of high art values and is taken to be part of the same phenomenon as mass humanity. There remains a strong sense of European elitism associated with Adorno and Horkheimer.

(4) A primary problem with such social-critical responses is what may be called the ‘ high altitude ’ or overgeneralization which is made with respect to both specific technologies, but also with respect to ensembles of technologies. ‘ Technology ’ remains a powerful, but nevertheless indirect and background phenomenon in the social-critical approaches.

A ‘ RADICAL ’ APPROACH

A third set of responses might be called radical. Here there are two forefront shifts regarding technologization which mark the more radical turn:

(1) Technologies are seen not to be neutral, but to be at least determinative enough to incline directions through their use. Types of technological determinism are not new in themselves, but the generalization concerning the non-neutrality of technologies is much more marked in this mid-to late-twentieth century interpretation. And, it is in this context, that Heidegger's analysis of technology becomes focally important.

Being And Time (1927) was noted above as containing a kind of proto-philosophy of technology in Heidegger's tool analysis. In the early context Heidegger applied a subtile phenomenological analysis to the human (*Dasein*) use of tools showing that technologies in use « withdraw » as objects of experience and become means through which *Dasein* relates to a world (*In der Welt sein*). Moreover, tools or technologies belong to a context or field of references thus revealing

---

[7] Cf. « De-romanticizing Heidegger », in Don IHDE, *Postphenomenology*, Northwestern University Press, 1993.

World. But it was not until the post-World War II period that a deeper non-neutrality of technology is shown.

(2) If, from early on, technologies show a world in a certain selective way, in the second move, ‘ technologies ’ become ‘ Technology ’ with a capital ‘ T ’. Technologies reveal a world – or at least a dimension of it – but Technology overall, in the later work, becomes identified with a general outcome of Western metaphysics. Technology, capitalized, becomes the characteristic of the late metaphysical tradition which employs « calculative reason », which « enframes » the entire world, and which « challenges » Nature, taking it to be « standing reserve » or a « resource well » [my translation of *Bestand*] within a framed [*Gestell*] totality. Here, most particularly in the 1954 « Die Frage nach dem Technik », but also in numerous other essays of the period, Heidegger *totalizes* his notion of Technology.

Nor does Heidegger stand alone within the radical approach. Jacques Ellul makes a similar move at the very same time. His *La Technique* (1954) takes technologization as the « wager » of the twentieth century. He does not, like Heidegger, metaphysicalize technology, but in another sense extends it into a totality in an even broader sense. « Techniques » suggest more than technologies in the ordinary sense – a « technique » is a way of doing things and thus in Ellul the totalization which takes shape is one which sees any ‘ rational ’ or ‘ calculative ’ or even ‘ quantitative ’ technique as a part of Technology. In the end one might say that there is nothing left outside technology in the century’s wager. Even ‘ Nature ’ is taken into technique *cum* technology.

Later, and in the same trajectory but from the perspective of the later Critical Theory School, Herbert Marcuse also totalizes Technology in *One Dimensional Man* (1964). Again, any analytic technique becomes a ‘ technology ’, such that even philosophy in its analytic form becomes technology. In Marcuse’s case, however, there remains an ‘ outside ’ which is a possible seed for revolt – but it must necessarily take the shape of rejection and a revival of the deeper (Freudian) desires.

(3) A third feature of the radical approach – but not necessary to its internal logic – is often the inclusion of a form of nostalgia or romanticism which can overlap with the also noted nostalgias of the Social-Critical Approach. The preference for hand tools, older technologies, and simpler technologies in Heidegger ; the nostalgia for

the ' natural ' in Ellul ; and the attempt to revive ' desire ' in an almost libertine fashion in Marcuse are indicators of this feature.

(4) The above noted features of the Radical Approach clearly have both conceptual and practical limitations. Conceptually, any philosophic move which totalizes a central metaphor risks the problem of having no means of counteracting the single inclusiveness – it functions as a kind of internal idealism for which there can be neither falsification nor counter-example.

(5) Practically, such approaches also risk such a high altitude analysis that all ' technologies ' are reduced to the same analysis. Thus any analysis of particular technologies is effectively precluded. Only the ' whole system ' of technologies can be analysed. Any single technology or system of technologies can become mere illustrations of Technology overall.

THE LATE TWENTIETH CENTURY

The above developments have taken the origins of philosophy of technology into the mid-twentieth century, into the sixties. It is clear that technologies, by this time, are increasingly drawing philosophical attention and response. But through the sixties the specific term, « philosophy of technology », is still not often used, even in its neo-Hegelian sense. The sixties, too, often were dominated by *negative* perspectives upon technology.

For example, although Jacques Ellul's *La Technique : L'Enjeu du siècle* was published in France in 1954, it was not translated to English until 1964 as *The Technological Society*. Along with Herbert Marcuse's *One Dimensional Man*, published the same year, and a plethora of related publications, the dystopian tone concerning technology dominated. This genre of philosophic critique of technology continued well into the seventies with a variety of authors including Hans Jonas, *Philosophical Essays : From Ancient Creed To Technological Man* (1974), Max Horkheimer, *Critique Of Instrumental Reason* (1974), and Langdon Winner, *Autonomous Technology : Technics-Out-Of-Control As A Theme Of Political Thought* (1977). This was enough to stimulate at least one humanist-engineer to rise to the defense of technology, Samuel Florman, who published a series of books attacking the dystopians. *The Existential Pleasures Of Engineering* (1976) and *Blaming Technology : The Irrational Search For Scapegoats* (1981) attempted to turn the tide.

Less obviously, however, there were groups of philosophers who, by the mid-sixties, began to realize that a *philosophy of technology* was called for, and that philosophy, if critical, should also be balanced. The journal, *Technology and Culture*, published the first collected set of essays seeking to develop this sub-speciality in an issue, « Toward a Philosophy of Technology », in 1966. And a new yearbook, *Research In Philosophy And Technology*, edited by Carl Mitcham, began publication in the 1970's. While the title of the yearbook self-consciously chose philosophy *and* technology rather than philosophy *of* technology in its title, it did begin to compile bibliographies and collections of essays on the « philosophy of technology » as early as 1973.

By the late seventies, not only had there been an explosion of collections, bibliographies, and articles by philosophers on technology, but books utilizing the title, « philosophy of technology », finally began to appear. Friedrich Rapp published *Analytische Technikphilosophie* in 1978 (English, *Analytical Philosophy Of Technology*, 1981), and Don Ihde, *Technics And Praxis: A Philosophy Of Technology* were among the first, the latter with the first English language philosophy of technology in a philosophy of science series (The Boston Series in the Philosophy of Science), 1979. The seventies also saw the beginnings of the institutionalization of philosophy of technology with a fledgling and sporadic set of meetings for a group which called itself « The Society for Philosophy and Technology », which began to meet in 1975 and finally organized formally in 1983. The eighties and nineties have continued the expansionist trajectory and philosophy of technology is today clearly recognized as yet another sub-specialization of philosophy. Introductions to the field have appeared in several languages, several book series continue, and one, the Indiana Series in the Philosophy of Technology, carries the specific philosophy of technology title[8]. And, with the aforementioned *Thinking Through Technology* (1994), there is even a *history* of the discipline. Thus by 1980, one can say that philosophy *of* technology has become a recognized field.

Also, by the eighties, one can detect yet another groundswell of atmospheric change. Totalistic and dystopian strands of technology critique begin to lose ground on two fronts. First, one can see today that much dystopian totalism is driven by skepticism about the centralist,

---

[8] Both *Research in Philosophy and Technology*, and a new series from the Society, *Philosophy and Technology*, continue.

industrialist, and « nature dominating » technologies of the late 19th through mid-twentieth centuries. Such technologies, forefronted by dystopians, I have suggested, were responses to concrete historical phenomena, particularly World Wars and the Industrial Revolution. And one of the engines which continued to power such interpretations relate to the technologies of the Cold War. Both military and industrial Big Technologies remained focal. However, with the collapse of the Socialist East, symbollized by the destruction of the Berlin Wall in 1989, different after-effects began to be felt. Centralist tendencies deteriorated into de-centralized and even regional revivals, the arms race – if not entirely stopped – at least slowed and downscaled to smaller weaponry, and the shock of newly revealed environmental disasters became of more import.

A second feature which affects the contemporary tone of technology critique may be associated with the proliferation of communications and particularly computer or *information* technologies. Such technologies no longer rely upon what could be called « energy metaphors » such as those developed earlier by Heidegger. « Information metaphors » are not the same as *standing reserve* or *Bestand*, but imply exchange, interaction, communication. The forefronting of such phenomena as the *internet* which is loose, decentralist, and interactive in present form, does not look at all like the secrecy-bound, centralist, and warlike expressions of cold war politics. Instead, the problems which emerge in the late twentieth century relate to the negativities which associate precisely with regional, ethnic, religious, and other downscaled phenomena in contrast to the large power blocs of the past. Insofar as philosophies of technology respond to and reflect upon historical phenomena, post-1989 history begins to look different.

These changes, not only were not predicted by the dystopian totalists, but it is hard to see how technological determinism could have foreseen such changes. Thus it can be argued that with the simultaneous demise of the Cold War, the shifting of focus to which « homogenic » actions have ' natural ' import. An instance here is the generation of greenhouse gases which are repeatedly characterized as approximately one-quarter of all heat retaining gases in the atmosphere. Similarly, Bruno Latour's suggestion that we need a third term, *hybrids*, of which he terms greenhouse gases one, is another attempt to overcome the nature/culture Modern distinction[9]. In short, unless we re-conceptualize

---

[9] Bruno LATOUR, *We Were Never Modern*, Harvard University Press, 1994.

our relations to the environment in such a way as to account for the interaction of human/non-human factors in a systematic way, we will end up reifying the modernist distinction between the human and extra-human dimensions of our world. Regarding technologies, in this context they are both ' hybrids ' and the mediators of the human/non-human contexts.

In the context of this problem set, there are also needs for analyses which can be applied and finely tuned enough to differentiate between types of technologies. Here the ' all-technologies-are-the-same ' approach of dystopian totalists clearly fail and thus can perform no normative role. A task for philosophy of technology is one which calls for careful, critical, and developmental analyses, not merely analyses which after-the-fact apply ethical band-aids. The role of critical philosophy of technology must relate to technology innovation and development as well as to result and effect.

Under the broad heading, environment, there are many sub-headings, all of which have strong environmental relations. Energy production (with nuclear power a serious late modern problem), remains important, as well as conservation strategies, corrective the forefront communications and information technologies, and the rise of cultural based types of self awareness, that the entire field of problems related to the social-political-and now cultural dimensions of the philosophy of technology, has changed.

TWENTY FIRST CENTURY PROBLEMS

Philosophers have barely begun to critically reflect upon what the forefront problems – particularly with respect to philosophy of technology will be for the twenty-first century which is about to dawn upon us. It can be argued, however, if the field is broadly construed, that there are three very deep and wide problem sets.

I. The Environment

The dystopian philosophers of technology often rightly targeted the Big Technologies of both industrial capitalism and industrial socialism with destructive trajectories regarding the environment. Those tendencies continue, although there has emerged a sub-set of environmentally protective concerns. While it may not be necessary to assay the full set of problems under this heading, there are several distinctly philosophical problems which may be suggested : a deep conceptual problem relating to the environment, is the very way in which it can be conceived. Here the controversies which arise are those which cut across much late twentieth century thinking. The ‘ Modern ’ set of distinctions which roughly or even sharply divides beings into the « natural » and the « cultural », no longer seems appropriate. Thus various ‘ postmodern ’ directions are being explored. For example, relatives of the « Gaia Hypothesis » are one such direction. The entire earth can be conceived of as an interactive system, one in development to restore previously polluted, depleted, or other ‘ natural ’ disasters caused by late modern Big Technology and Industry, all fall under this heading.

II. Multiculturalism

A second region which has been forefronted in both the problem and possibility area, is the emergence of contemporary forms of multiculturalism. The communications and image technologies have opened the entire human community to interaction. Combined with travel technologies, the virtually constant intermixture and interconnection between cultural groups now characterizes the texture of late modern life.

The negative reactions, exemplified in virulent ethnic revivals and conflicts, are but the negative side of what could be called postmodern pluriculture in which cultural borrowings, exchanges, and mixtures increasingly describe actual contemporary multiculturalism. And while most obvious within the international youth cultures, such eclecticism increasingly marks larger segments of traditional cultures as well. Here the technological components are those of travel, communications, and information technologies which are the media by which the exchanges occur. The forms of life which are eventuating need not only critical analysis, but explorations of the possibilities which lie within them. On the political and ethical front, a re-examination of neo-enlightenment policies should be undertaken with respect to the values of tolerance, forebearance, and restraint.

III. The Micro (Bio-and Nano-technologies)

Modern science, instrumentally embodied, has led to the human capacity to manipulate the micro-aspects of the world at the level of atoms in the physical world, and genes in the biological domain. And while such micro-technologies stand at the extreme end of the spectrum from Big Technology, the potential here is equally or even greater than the macro-technologies of the 19th and 20th centuries. These are the frontier technological developments which already are detracting from the past dominance of the physical sciences. The biological sciences, already funded in the billion-dollar ranges, are rapidly becoming forefront technological domains. Similarly, for philosophers of technology, these domains call for careful study, analysis and critique.

ECONOMICS

Vittorio MATHIEU
University of Turin

1. THE OBJECT OF THIS SCIENCE

Robinson Crusoe has just landed on the uninhabited island, and he is immediately confronted with that sort of problems that are called by economists « best allocation of scarce resources ». These resources included : (1) himself, as a skillful seaman ; (2) the boat and some tools saved from the wreck ; (3) what could be exploited on the uninhabited island. The three items correspond to the three fundamental factors of production : labour, capital, land ; and Robinson must combine them in such a way, that the result be the best. Therefore Robinson Crusoe has been taken numberless times, as a paradigm of the *homo oeconomicus*.

Nevertheless, this paradigm is defective : the island is uninhabited and Robinson has no occasion to *bargain* for anything. Economic activity is not conceivable without bargaining. Of course Robinson can compare the benefit expected from a certain behaviour with the benefit he could take from a different behaviour, and in this sense, he bargains with himself. For instance, it is advantageous for him to abstain from plucking fruits for a day, in order to prepare a fish trap that shall free him for a long time from the necessity of fishing with hook and line. Robinson « economizes » his efforts. He chooses a capital increase instead of enjoying immediately its fruits. Nevertheless, to apply to his conduct the theoremes of economical science would be a strained misrepresentation, because the best allocation of scarce resources can be persued without any reference to a *political* (i.e., social) economy.

The same can be seen through another paradigm proposed by Paul Samuelson in the first editions of his textbook. A hive of bees solves many problems concerning the best allocation of resources. For instance,

P. Kemp (ed.), World and Worldhood, 109-147.

it chooses the best form of cells for keeping the maximum of honey with a minimum of wax : this form is the hexagon, and a rigorous mathematical calculus gives the same result.

Samuelson insists : « Any society, whether it consists of a totally collectivised communist state [...], a capitalistic industrial nation, a Swiss family or Robinson Crusoe, and one might almost add a colony of bees, must somehow meet three fundamental economic problems :

1) what commodities should be produced and in what quantities ;

2) how shall they be produced ?

3) for whom are they to be produced ? ».

And Samuelson concludes : « In the bees colony all such problems, involving an extraordinary elaborated cooperative division of labour, are solved automatically by means of so called biological instincts ». And yet economy could hardly progress through the study of bees and, besides that, political economy gives no acceptable interpretation of the behaviour of the hive. Unlike Robinson in solitude, bees actually have a social behaviour, but the result cannot be considered as an economical fact, because no bargain occurs. (That is the reason, perhaps, why bees disappear in later editions of Samuelson's textbook).

In my opinion, the economical conduct is but the second (or third) best solution of the problems Samuelson points out : the first one is biological instinct (of the bees) ; the second is *love*, always defective and nevertheless effective in a family or in a convent. And we can add a fourth solution, largely spread among human beings : slavery. The economic solution distinguishes itself from all these different solutions of the same problems, because it consists of a relatively free bargaining.

2. AGREEMENT V. CONSTRAINT

To understand what economy is, we must put together all the faulty suggestions, and we obtain : (1) the facts taken into account by political economy are consciously deliberated actions, aroused by a desired outcome and not by a biological instinct ; (2) this result can be reached only with the cooperation of other people, whose assent has to be voluntary ; (3) the « aggregate » result, economy is interested in, is not the same that individual contractors had in mind.

Point (3) explains why bees can give a contribution. Insect societies do not need to deliberate about their future : their instinct, matching spontaneously with other facts of nature, shall produce congruous results. Human societies do not enjoy the same advantage, but precisely in economic activity the aggregate result of indivual actions gives place to a spontaneous outcome nobody looked for, as if it were the result of an instinctual behaviour.

This is the role of the bees in Mandeville's fable, that is precisely the conceptual sources of the science called « political economy ». Adam Smith was not particularly soft towards Mandeville, because he was impressed by a certain amoralism of his doctrine. And, in fact, the apology of luxury in Mandeville's apologue is neither new nor convincing. That the satisfaction of a luxurious mood give employment to people, who otherwise would remain unemployed, is no argument at all : people shall have no trouble in directing their efforts towards better aims. But, what is absolutely definitive for the birth of political economy, in Mandeville's essay, is that the aims of the single bees are quite unlike, and even divergent, from the well-being of the whole hive, apart from the opportunity of opposing them, as « private *vices* », to « public virtues ».

Of course, this divergence is not at all the case of the bees : it is the case of human beings. What Mandeville points out is that a spontaneous outcoming comes out of a human conduct, *although* it is human, *viz.* deliberate. And this outcoming is precisely the object of political economy : the global result of a split set of resolutions has an independent sense, that none of the deciding subjects had in mind. That an instinctive conduct has results that no individual insect had in mind is obvious, but the case of political economy, on the contrary, is astonishing, because human beings *have* something in mind, and this is not the global result that will ensue. Political economy takes the place of the providence in an age of secularization.

If we suppose that a « strong man », in a planified society, is really strong enough to give an answer of his own to the three questions of §1, this would be a matter of ordinary technique. No special science called « political economy » would be needed to explain the success. But, on the contrary, we observe that the strong man generally reaches results awfully distant from his desires, no matter how skillful and well-meaning he is, and this « vengeance of the economy » is what we really must explain, as well as the unpredicatable outcoming of individual activities in a non-planified society.

But is not the whole human history an « aggregate » result of voluntary actions, none of which had that result for its aim ? Yes, it is, precisely because that history is human, i.e., made by human beings, acting egoistically like *homines oeconomici*. Without the behaviour political economy is concerned with, there would be no history at all. (At the end of Marxian eschatological utopia we find, in fact, the end of human history).

3. OMEOSTATIC EFFECT OF THE INFORMATION

Let us summarize some conclusions. (1) Economical actions are, in the proper sense, *transactions* : operations in which two or more subjects, persuing different and even contrasting purposes, *agree* in some decisions. (2) These purposes meet in a common result. (3) This encounter is what we call with the latin word « convenience », that means : two or more persons *meet* in the same place. (4) This meeting for convenience must be spontaneous and (taking « convenience » in a new sense) grounded on the research of a selfish advantage. (5) Although the behaviour of each party in these transactions is guided exclusively by a selfish intention, the whole outcoming presents the society with many advantages, that other types of answers to the questions put in §1 (instinct, slavery, love) would not give. (6) There is a gap between the level of the individual purposes and the level of the aggregate outcoming, that no planified action is able to overcome.

Let us give an example. Two persons are engaged in the most simple economical transaction, a sale, and both the buyer and the seller are interested exclusively in obtaining the best conditions, each for himself. If they are bargaining a little stock of tomatoes, they are not even touched by the idea that their business will result in an imperfect, but effective adjustement of the quantity of tomatoes produced all over the world to the wants of the world population ; moreover, that this fact will influence the allocation of resources in the whole agricultural field. Nevertheless, the countless number of similar events, nobody knows anything of, has exactly that result of fixing and « equilibrium price ». If you try to do the same through a scientific analysis of the situation, you shall be inevitably unsuccessful. This is highly paradoxical : if nobody knows, the adjustment will occur (albeit imperfect), if you consciously plan, a disaster will be the result.

Only after nearly two centuries of economical studies the real function of the market was discovered : to give information. But what sort of information ? Just a statistical information you are compelled to « realize » in your conduct, more than in your conscious knowledge. All specultative enrichments are based on *news* you have, and other people have not ; but the process can develop in two ways. You can find out, for instance, that the managing director of a company has crazily fallen in love with a vamp : you sell a number of shares of that company (without holding them) and after a month or two you cash the difference. Or you can foresee e.g. that a lunatic mood will induce women to prefer next year either terribly long or awfully short skirts, and you strike off every skirt of reasonable length from your production. This is the result of an impression, more than a knowledge, and the information you get from the market is generally of this kind.

We have intentionally chosen our example in the fashions, because one could object that a preference can be *artificially* induced. Nevertheless, the starting point is always spontaneous, as an incoming trend, of which the situation gives at first very slight hints. The winning businessman catches these hints before the others get aware of them, like a peasant who is able to foresee the rain of tomorrow, better than an expert of meteorology.

The parallel with meteorology enlightens us about the statistical character of market information, but it leaves out what is more important : the *omeostatic* effect on the market itself. For that, it is not necessary to foresee the future : the microeconomical hints are sufficient to produce a natural feedback, without any intervening deliberation. The rising price of a commodity can be analysed by the board of directors of a firm in order to decide how much the production should be increased, but before that, it induces the simple buyer to look for another commodity that can replace the first. What is important to point out, in any case, is that at the end the intention, both of the buyer and of the board of directors, concur to the same result : to adequate quantities and needs and to reduce the losses. Precisely because their aims are egoistic, the result is advantageous for the community (as Mandeville said).

4. THE « INVISIBLE HAND »

We meet therefore in economy *two levels* of events. On the lower one, a crowd of individuals strive to obtain a profit, conveying their decisions towards transactions on which other subjects agree, out of convenience. Then, on the higher level, a set of adjustments results, nobody was aiming at, as if an immanent providence were their source. Sometimes the result can also be unfavourable, or can be prevented by intentional measures that have not an economical source, particulary for avoiding the so-called externalities. But a lot of effects are favourable without being intentionally aimed at. This secular providence is the reason why the newly born science of political economy struk so favourably Hegel, who saw in it an analogous of his own « cunning of the Reason » (*List der Vernunft*).

This difference of level has given place to a distinction between two main branches of the economical science : micro- and macro-economics. The difference does not depend on the importance of the deliberations at stake, but on the type of their effectiveness. Microeconomics reflects the individual decisions taken by individual operators. Macroeconomics reflects the effects on the economy of laws, or law similar measures, concerning a whole class of events. These measures are generally taken by public authorities, but this is not the point : when a State buys or sells an estate, it is acting as a private operator and the event falls within the competence of microeconomics. Only when it acts as a public power the effects are of the macroeconomical type.

Both aspects of economical science belong strictly together, and in fact Smith's famous treaty refers in its title to macroeconomics, because it deals with the richness of *nations*, but the content insists mainly on the behaviour of individuals. Presently we have handbooks which deal separately with the two aspects, and in many textbooks the section on macroeconomics precedes the section on micro-. In my book on the philosophy of money, I pointed out that this is an ideological choice, and the reason of this choice is easily understood : the author wants to emphasize the importance of political interventions. Against this ideological choice I argued that an economist should take the opposite ideological choice : microeconomics first. Antonio Martino (later a Minister of Foreign Affairs) rebuked me, arguing that only the former choice is ideological, the latter being not ideological at all. I find now that he was right, because the order of priority is founded on the nature of the matter, i.e. on the fact that *individual* decisions have macroeconomical consequences as a result, and only in this case we can speak of economy. Of course, also the opposite way can be trod : tax legislation influences nearly all microeconomical decisions and so on ; but this pertains to politics as such, not to economics. This introduces a distinction that, from my viewpoint, is capable of clarifying many misunderstandings : the distinction between economical facts and facts of other nature that have economical consequences or effects.

5. JUST ONE ECONOMY, MANY WAYS OF PRODUCTION

Let us take, as an example, a theft. A theft has no doubt economical consequences, if it is successful : it makes somebody richer and somebody else poorer. But it is not an economical fact, because it lacks that agreement or reciprocal convenience which is the most important feature of an economical transaction. The same must be said of many measures that are perfectly lawful, like taxation. Theoretically, in constitutional States, you could conceive taxation as an economical transaction, i.e. as the result of a bargaining between the executive power and the representatives of tax-payers : citizens receive some services and must pay for them (« no taxation without representation »). In non-constitutional States, this is not the case. An elegant question arises, in my opinion, if you ask, whether the fixation of the discount rate by a Central Bank is an economical fact or a political fact with economical consequences. If the Bank of issue were a private company – as it often pretends to be – the discount rate would be the price for the transformation in time of a sum expressed in the currency issued by the Bank, and the customers could bargain for it. But, on the other hand, the customers of such Banks are generally constrained customers and have scarce opportunities to obtain a hearing, while the main customer, the State, is in fact able to command the rate it wants.

The distinction we have spoken of is important, because, if we remove from the economical science all arbitrary constraints from outside, there is no more reason for admitting *two* (or even many) types of economy, namely the market economy and a planned or socialistic economy. Every economical action is planned, of course, but it is planned on the lower level by individual actors (also a corporation or a public body is individual, when planning transactions). Any economy is market economy and « market » is a redundant word just expressing the ideal place (the market place or *agora*) where freely taken decisions meet. No different type of economy exists. There are of course different types of *production*, not related to economical production : mainly the slavish production, which does not arise from a bargaining between master and slave. Some factors of this production have an economical origin, e.g. when the slaves are bought. And even the constitution of slavery can be regarded as a « quasi contract », because the man who surrenders gives up his freedom for his life, but this is a borderline case of economic transaction.

In a socialist economy there is always a component of public constraint, more important than in a liberal one. The only owner of the main means of production being the State, the directors of a firm are obliged to follow the decisions of the ministers, otherwise they could be executed. But, in general, it is nearly impossible to organize a real production only on this basis, avoiding any economical transaction. You cannot obtain an appreciable efficiency from the staff without agreeing with each component the task he has to fulfil and the wages he has to receive. Beside that, both in socialistic and in liberal economies there is a big submerged part, which escapes the authoritative dispositions. This does not mean, in any case, that these dispositions have no economic effect, but they do not pertain to economics. If the power establishes a political price for a commodity, the market price – that is the only economical price – of this commodity will rise. If the power sentences to death everybody convicted of black-marketing, the price will rise even more, but it remains a *price* balancing supply and demand, what is not the case of the so-called political price. What about the transactions performed under the imposed price ? Are they economical or not ? Of course, they are. But either the buyer or the seller takes into account circumstances which have nothing to do with a free choice : death sentences and so on.

This leads us to understand in which sense an economic system can be « mixed ». In fact any economic system *is* mixed, but mixed with decisions of a different sort. The mixture does not occur between two different types of economy – as if there were a market economy and a socialist one –, it is mixed between economical and non economical decisions (unless, of course, you agree with Benedetto Croce, for whom *any* human decision is an economical one).

This conclusion depends upon the rigorous definition of economics we have proposed. If you choose the definition : « a science aiming at the best allocation of scarce resources » there are of course many possible procedures to reach this end. But if you add that the economical allocation must result from a « meeting for convenience » of free operators, then there will be but *one* economy, under any political climate of the world.

6. HOW LARGE THE SHARE OF ECONOMY ?

The discussion between liberals and socialists about the policy to choose is perfectly admissible, but it does not concern which type of economy should be chosen : it concerns how large a *proportion* should be given to economy in human affairs. The breeding of the children, for instance, is generally performed by parents, without any economical transaction, but it can also be assigned to a State orphanage, or to a wet-nurse. In the last case, if the nurse is not bound to the family by a bond of slavery, the breeding will be an object of bargaining. In some societies, particularly in the so-called « hydraulic empires », that frequently developed at the dawn of history, the part reserved to economy was indeed very small. On the contrary, in highly developed societies, like the European society of the XIXth century, this part was exceptionally large. Then, worries arose about leaving such a huge space to the free decisions of private individuals, where the rich easily prevailed on the poor. In the Incas empire, the decision to destine a daughter for prostitution always, or nearly always, depended on the connection of the family with public powers ; in our societies, on the contrary, it depends on economical considerations. In view of changes of the sort, particularly some utopian socialists devoted themselves to thinking out a new social order, in which absolutely no relation ought to be of the economical type. The simplest device conceived for this purpose (for a reason we shall shortly see) was the abolition of money. The failure was complete. The officially abolished money showed an irresistible tendency to reappear in disguised forms. We are then authorised to conclude, not only that economical behaviour is a peculiar feature of the human kind, but that it can by no means be wiped out of a human society.

7. ABSTRACTION

The continuous interference of other classes of decisions with economical decisions produces that classical character of the (pure) economical science which is called « abstractedness ». In other words, political economy is in any case obliged to leave out factors, which in fact influence the actual situation. The interest of our science lies precisely in this possibility and necessity of being abstract and of leaving out important features of the concrete human conduct. Many many times I have heard or read this remark : « The *homo oeconomicus* is an abstraction ». This statement is absolutely right, but one can doubt its usefulness, because I never heard anybody maintain that the *homo oeconomicus* is not an abstraction.

The possibility to reach an effective abstraction in a matter as confused as bargaining among millions of people is, as such (*an und für sich*), very impressive. It enables to extrapolate some regularities, which assume the showy name of « laws », distinguished from one another by their object or by the discoverer's name. There is, for instance, a « law of diminishing returns », owing to which, if I eat a partridge a day for a year, I'll become sick and tired of it.

The most important performance reached by abstraction was the mathematical theory of *general equilibrium*, achieved by Walras. Equilibrium is a situation which never occurs, it is only approached. But it is just this precarious and approximate approach what makes equilibrium so interesting, from a scientific point of view. Suppose to see a stone in an archaeological park, and to see it again five years later : you shall probably not be surprised at all to recognise *the same* stone. But if you consider that that piece of stone is constituted by an astounding number of particles which move frantically, as far from one another (respectively to their size) as the earth and the moon, the fact of recognizing the *same* stone five years later becomes an interesting subject of reflection. Well, the catallactic activity of some billions people, each interested in his selfish presumed profit, seems the most disordered condition of things : everyone follows different subjective impulses, whose causes are often unknown to himself. That such a situation tends to an equilibrium – not static, but perfectly determined as a stationary state – is even more interesting than a physical equilibrium, because the forces active in physics (or « interactions ») are knowable with appreciable precision, while the decisions taken by an entity endowed of free will are strongly influenced by whimsical and erratic fancies.

The most simple and comprehensible equilibrium in political economy is the equilibrium between supply and demand of a single commodity, connected with the price. Exchange quotations indicate the price at which, at a given moment, the offer of a security exactly equals the quantity of the same security wanted by the buyers. Outside the stock-exchange, the adaptation is not so rapid, but any positive or negative variation of the price causes a feed-back, both in the production of the commodity and in the demand. What is more important, is that the adjustement affects the price, not only of that single commodity, but also of *all* commodities that can in any way become a substitute for it. Walras's model and in general the marginalist theories have mastered skilfully such situations, and the fact that they deal with abstraction makes even more impressive their capacity of prediction.

8. DEDUCTIVE SYSTEM

And yet the science of economy was already sentenced to death, some years before Walras's systematisation had begun. On occasion of the centenary of Smith's *Wealth of Nation*, a distinguished economist proposed to abolish the section for economy in all scientific societies, arguing that all was to be discovered in that discipline had been already discovered by that time. The same opinion was expressed by an eminent physicist about physics, at the end of the XIXth century, but the case of economy seems more serious, in consequence of the formal and nearly hypothetical character of this science. More than to physics, economics is similar to the *theoretical mechanics* of the XVIIIth century. Its investigations are not about the real world, but about a hypothetical-deductive system : *if* a population of intelligent beings interacts by way of bargain, in order to fulfil its wishes, *then* such and such regularities appear. What is really acting in this system, what is the meaning of the hypothesis, that human beings have preferences and can *decide* according to them, falls outside the economical investigation, and perhaps outside any scientific enquiry at all. The question involves even the notion of « free will » that always produced harsh desagreements among philosophers and theologians. Economics simply assumes : that human beings have a natural propensity towards their own well-being, and that they are capable of determining their choices in view of that end. Whether this assumption is right or not, and whether this or that single choice is really appropriate, is immaterial. What matters is the statistical result of their behaviour.

Its abstract character and the relative narrowness of field do not deprive economy of its enormous human and scientific interest ; but it is, of course, a deep inconvenience for people who intend to devote their life to that science. Shall they remain bereft of their job ? In order to avoid this issue, the greatest part of economic enquiries today is devoted to the influence that non-economical (mainly political) decisions have on the economical life. This has brought about a school of economic thought, called « institutionalism ». Thanks to that device, it is possible to hand out each year a Nobel Prize for Economics (or even two, if the prize is divided), notwithstanding the 1877 sentence to death.

Also decisions of a non-institutional source may be taken into account. E. g., taking the cue from the existence of an enormous liquidity in some bank-books of Rome and Palermo, I remarked that the criminal activity has a part in the well-being of a currency, because it increases the demand of money (owing to the wise principle of criminality : « *touchez pas au grisbi* »). Inversely, the criminal behaviour has been studied in USA from the strictly economical point of view of the ratio costs-benefits. When the costs and the risks (in this case not insurable) grow, also the profit must grow, or the activity shall be abandoned. The criminals become « honest », because they find it no longer convenient to break the law. Legal severity appears then as an economical device for lessening the « marginal ophelimity » of the crime. (This approach was present in the USA jurisdiction before it was theoretically announced : the famous gangster Al Capone, who was responsible for dozens of murders, was finally found guilty only of tax-evasion : kill as many persons as you want, provided you pay a sufficiently high tax).

9. IMPOSSIBLE PREDICTIONS

The formal character we have spoken of explains the gap existing between the geometrical previsions of the economical science and the substantial but uncertain foretelling of economical operators. The scientist says that (with other conditions remaining the same) an increase both of the probability and of the importance of a penalty will obtain a drop of the criminal activity, but he is not able to say *when* and how much. For this he should introduce into his models some variable of psychological nature – e.g. « propensity to break the law », and so on – which are quantifiable only *a posteriori* (like the « propensity to save » and other formulas of this type, which appear particularly in the followers of Marshall's school). In fact, propensities and similar psychological issues lie completely outside the economical science, for which only the behavioural consequences are important. The strength of a propensity (to work, to enjoy free time, to eat a cake and so on) is known only by the effect and cannot be used to foretell it. This is the reason of the essential default of the scientific previsions in economy : the inability to say *when* and *how much.* The prevision can just enter as a general advice in the blue prints of the operator, but not as an element of calculus, like the prevision of a physicist. If it were not so, the best economists would rapidly become the richest people in the world. It would be sufficient to be able to say (not *that*, but) *in which moment* a price at the stock-exchange shall undergo a variation for getting enormously rich in a short time. No economic knowledge is able to do it. Lucky speculators belong rather to the class of weather-tellers in economy, with little scientific preparation, but with « good nose » for business.

Somebody could argue that this is due only to an imperfect analysis of each individual situation ; or to the fact that the final decisions of each operator cannot be known in advance. But the inconvenience does not lie in the unpredictable free-will. Recent considerations of the science of « deterministic chaos » show that there are problems which cannot be treated (in a « polinomial time » with the number of data), because the number of final possible situations increases exponentially with the changing of control parameters, which produces « bifurcations » in the path, independently from any free will. In a computer, whose computational unity of time is a millionth of a second (very speedy, indeed), the computation time for hundred bifurcations is 2^{100} seconds = 10^{20} years, more than the age of our universe. The new frontiers of statistical science explain why political economy does not allow detailed previsions, irrespectively of the source of bifurcation : whether they are produced by really free actors, or by very slight impulses from the surroundings. Free will is a presupposition of political economy, because this science deals with « decisions about the best allocation of resources », but what these decisions are, metaphysically, lies outside the theorems which constitute the body of the science.

10. THE ROLE OF MONEY

In view of a business, the operator has to judge if the transaction is convenient for him or not ; and the same is done by his counterpart. When eventually both agree, the exchange occurs, and the exchanged goods (or services) prove thereby to have the same value. Nevertheless both parts are satisfied with a gain. This difference between subjective and objective judgment is the first hint of the duplicity of level, introduced by the « invisible hand » between individual actions and economic outcome.

To evaluate is : measuring the value of a thing in terms of another thing, used as a standard ; to judge, for instance, how many pounds of silver is worth a pound of gold. But generally a third thing is used to compare both sides of the identity. E.g. the poet of *Iliad* estimates that Diomedes' armour was worth hundred bulls, Glaucus' armour nine bulls. Bulls are used as a standard of measure, for expressing the price, or the relative value of two armours of different material. We can say that in this case bulls are « money », but in a particular meaning of the word : money as *pecunia*, hardly distinguished in English from money as *moneta* (*nummus*). *Pecunia* could be translated with « wealth » ; but, while Latin languages generally distinguish between *argent* and *monnaie,* in the German languages there is just one expression for both : in German *Geld* (= value), in English *money* (= coin). Bulls were scarcely used in archaic Greece as a medium of exchange : for this purpose the Greeks disposed more frequently of differents gadgets : spits, tripods, or even slaves, according to the importance of the object. Primitive Greeks had different standards for the three main functions of money : (a) measure of value ; (b) purchasing power ; (c) store of value. All these functions are fulfilled by our currencies, and *anything*, generally speaking, can fulfil these three functions, but not with the same degree of suitability. Two slaves, for instance, can involve tremendous differences of value, much more than two bulls ; before accepting a slave for payment, you will probably submit him to an attentive examination, and you cannot use slaves as a general standard of value. A rare stamp is an excellent store of value, but it is not a « legal tender », and it is not practical as liquidity.

What I want to point out is that, although money in sense (b) appears relatively late in history, money in sense (a) is as old as the economical behaviour, and consequently, as old as mankind. Somebody maintains that the *barter* precedes the exchange through the intermediation of money. This can only be of money in sense (b). If barter is an exchange decided on the basis of convenience, it presupposes a common standard of evaluation, i.e. money in sense (a). How could you otherwise judge whether the exchange is convenient or not ? Of course, not all exchanges occur on this basis. A big hefty fellow can compel me to exchange my Rolex with his Seiko by very different arguments ; but this has nothing to do with barter. Also the exchange of the armours between the two Homeric heroes was meant as a mark of friendship, and involved no money. But the poet, living in a period when the concern with profit had largely developed, comments that Zeus had deprived Diomedes of his whits, precisely because he had omitted any economical evaluation.

The meaning of the passage is : *if* the two heroes had taken the economic value into account, *then* they would have been obliged to express the value of the armours through a common standard : e.g. bulls. In fact, no concrete token is needed for this purpose : we could use just two numbers, 100 and 9, and write the proportion : Diomedes' armour/Glaucus' armour = 100/9. In this second ratio all tangible objects have disappeared. What would be the inconvenience ? Not a merely practical one. When two single things are compared, each of them can act as a standard of value for the other. But when *all* commodities and services are compared, the complication would become untenable.

We could imagine an immense computer where the relative value of everything (including obligations) is registred ; so you could suppress all reference to a given token and introduce a universal clearing. You get a dinner in a restaurant, and instead of giving the equivalent of a glass of wine as a tip to the waiter, you let inscribe all these variations in your asset, and in the asset of the waiter and of the restaurant, in the computer's memory. But why is it impossible to resort exclusively to this electronic money consisting of numbers ? Not only owing to the limited capacity of those machines. An enormously powerful machine could give a picture of the instant situation of all relations of debt and credit, and receive the news of their variations, but it could not calculate the variations in all relative values that a single transaction *is going* to produce, no matter how small, combined with all other transactions performed everywhere in the world. This is a matter of *evaluation* on the side of the operators, who have to *decide* in view of the future. The machine can only register *later* the effects of their decisions, but what gives reasons for deciding is only the market, by means of the prices. These are at the same time the result of all past decisions and the hints to decide again. They express, in other words, the *passage from the past to the future*, while the computer depicts only the static situation of the past. Prices are not scalar quantities, they are *vectors*, because they are always directed towards something, and their direction changes according to the global movement of the market.

11. MONEY, NOT A COMMODITY

The circumstance we have spoken of gives also the reason of the passage from money as standard of value to money as purchasing means. The reason generally given is that the exchange of two commodities is usually made easier by means of a third, which must have following properties : (1) to be useful for everybody, and therefore universally accepted ; (2) to preserve its shape for a long time ; (3) to be as divisible as possible ; (4) to be easily conveyable. As long as only two commodities are exchanged on a market, for instance fish and corn, a barter can take place without money in sense (b), the relative price varying according to season, fishing opportunities and so on. But if the operators have to buy also nets, boats, ploughs and so on, it would be awkwardly unconfortable to receive a big quantity of fish for a boat, hoping that somebody shall be willing to accept this fish for a plough and so on. Then, obviously, corn will undertake the function of money as a universal means of exchange, and therefore also as a standard of value (*numerarius*).

This fact gives rise to the mistake of confusing money with a particular *commodity*, and the sale with a delayed double barter. (Money can be conceived also as a non-existing commodity, like Sraffa's money, but it remains nontheless a commodity). This reduces the importance of money in economy and conceals its peculiar character, which no commodity possesses. Any physical object can assume the function of money, but it looses its physical properties as soon as it assumes that function. We have just seen it in the case of the Homeric bulls : they do not intervene as real beasts, when they are adopted as a unity of measure, and could be replaced by anything else, also by a number.

At the same time this number must be qualified by an arbitrary *token*, in order to make the objects comparable, not only with one another, but with anything else. When you have got this possibility of universal comparison, you can suppose to change this X, called money, with all goods and services you want, because at this point you have reduced the token, you have started from, to a *symbol* : that is to a token which has the property of συν-βάλλειν, of « putting together ». As everybody knows, a symbol is a broken bit of solid matter : if the two pieces combine perfectly, their common origin becomes evident (it would be impossible to produce a perfectly corresponding fraction in another bit of material). This is also the possible origin of a banknote :

you tear away from the page of a counterfoil-register many pieces of paper, each with a corresponding progressive number ; when the ticket is presented, you are able to recognise its origin. Meanwhile, the note can circulate as widely as you want. The same proceeding was employed to « accredit » a messenger, to recognise a nephew, and so on. This possible recognition gives to the holder a *credit* of a particular kind, and we want now just to discover of which kind is the credit given by the money to its bearer.

Now, if money is a symbol, the real support is unavoidable to supply the *body* to be split, but what matters is not this body, but the *cleaving* as such. Later on, you can replace this sign by another sign of origin : coinage, printing and so on. These signs have again a physical support, which keeps also a value as a commodity : banknotes, e.g., can be used to paper the walls, but this has nothing to do with money. The monetary employ of an object (like gold, shells and so on) can remain *beside* its employ as a commodity, but they must be strictly discriminated, if you wish to catch the peculiar function of money in economy (and in the whole of human life).

12. MONEY « CONDENSES » THE TIME

As we have seen, the role of a symbol is to *displace* a credit : it is the recognition that something is due to the bearer (it can be : attention to what he says, a service to his person, and so on). But a symbol works specifically as money when : (a) the identity of the bearer becomes unimportant ; (b) the contents of the credit become indefinite (only its extension being determined : « how much », not « what ») ; (c) the debtor becomes unknown : no matter who (s)he is.

Let us explore the degrees of this process through an example. When the Italian State was horribly lazy in coining, our bars were embarrassed in giving the change for a coffee : the casher prepared for this purpose a stock of bon-bons. I did not like bon-bons, so I converted them immediately into a tip for the barmen. Other customers did the same. Before leaving the duty, the barmen returned the sweets to the cash, in exchange for a little banknote. The character both of the symbol and of the actors was not sufficiently generic to give to bon-bons a wider circulation, but they had really acquired the role of *money*.

Now a counter-example. In Clair's *Dernier millionnaire*, when the money disappears in Casinaria, a drink is paid with a cock, the balance is a chick and the tip to the waiter an egg. Why do we laugh ? Or when at the roulette the would-be suicide drops his Smith&Wesson on the number that comes out, and receives 36 Smith&Wesson, why do we find this absurd ? Because the content of the symbol is not sufficiently *abstracted* to represent a genuine monetary credit : it maintains the nature of a commodity.

Now, what happens when a symbol becomes money ? What is the content of that credit, symbolized by a coin, a banknote or a cheque ? It is a performance, that somebody will carry out for you, nobody knows who, and nobody knows what. But how to be sure that such an indeterminate obligation shall be fulfilled ? No expectation of this kind is safe, and nevertheless anybody is ready to convert his expectations into money, and wait. Money is an embodied expectation : shells, mats, coconuts, any sort of ornaments ; cigarettes in concentration camps, and so on. Those symbols are accepted in change of a service, because the receiver expects that they will be welcome again, in change of another service ; and so on endlessly (or at least as long as that specific currency fits with the cultural surroundings).

The fundamentals of these expectations shall be treated afterwards. At present we must remark an important feature of money : money allows a displacement of expectations in *space*, but mainly it allows a displacement also in *time*, towards the future. It is apt to delay decisions. It allows to decide what to do later, and to whom the symbol has to be addressed. This property is decisive, because only if a decision is *delayed* (at least for a moment) it can be technically free, namely a *choice* and part of a *project*. Otherwise you answer the situation automatically. Now, economy is an encounter of free decisions about what is convenient : so without the delay allowed by the money no economic activity would develop.

The conclusion is that, among the three main functions of money listed at §10, the most important is the *third* : store of value. Thanks to this property money can fulfil also the other task : to intermediate buying and selling, to deliver from obligations and, of course, to serve as a rod for measuring the wealth it « contains ». After all, the first idea you get into your head when speaking of money is precisely the idea of a *mastership on somebody* (you do not yet know) for obtaining from him something you want. And that is the idea of richness.

13. IDEAL NATURE OF MONEY

In fact, how can you benefit from your money ? Only *handing it out* to somebody else. To enjoy a good, you must appropriate it to yourself : to enjoy your money you must do the contrary. (Therefore I said that money is the reverse of a commodity). But what can you expect from other people in exchange for your money ? Either a product or a service. A product is in turn the produce of services, so we can say that what we expect from money is *labour* : either a precedent labour embodied in an object, or an actual work, performed by somebody ; generally both. In other words, money is *energy*, or potential labour, ready to transform itself into an actual and embodied work. Therefore it is symmetrical to all other forms of wealth. Thanks to its indetermination, it is freedom : freedom in respect to any *concrete* object of wealth, and freedom from social constraints.

This explains the greed for money, remarkable in all developed societies. The miser is so reluctant to restrict the choices that money ensures, that he declines to enjoy its richness even forever. But also normal people generally prefer to receive than to give money ; and the seller thanks the buyer, not the other way round. There is then equivalence in quantity, not in quality ; liquid assets are preferred, because they allow immediately a choice. What we have said about buyer and seller is not valid, perhaps, when big estates are exchanged against money. Here the property is conceived as an « activity », not as a potential energy against an actual one. This is the very difference between sale and barter. In fact very big exchanges approach often the barter : the goods (for instance, football champions) are evaluated, but just the balance is paid in money. Barter is hardly a primitive form of exchange, it takes place more frequently among nations, or very big corporations, by way of a clearing. It is not primitive, because it presupposes the sale, or, better, two mental sales ; a passage from actual to actual, through an evaluation, and therefore through an *idealization* of what is actual into the representation of a potential activity.

The *ideal* nature of money is what I meant saying that money is the opposite of a commodity ; and in my view this is important, because it expresses the direction of the money *towards the future*. Ideal, in this case, has no Platonic meaning, it means simply that money is not an object, but an expectation.

We meet here with a curious overturn of meaning : when we exchange a real good v. money, we say that we « realize » it, instead of saying that we « idealize » it. This happens because reality, in economy, is the freedom of choice : a void, in which we trust to attract other people to « convene » with us. In respect of the reality, this possibility is indeed a void, and to confuse this void with a *thing* (as it happens when we speak of money as if it were a mass) is the biggest naturalistic fallacy of the whole economics. It involves many serious mistakes, both theoretical and practical.

But, if money is a void, how can it be filled ? In other words, what gives to the value of money its consistency ? Money has a value because it is accepted, of course, but it is accepted as far as we think that it has a value : can this circle not be a vicious one ?

A commodity money seems to solve the problem, because it has also a « value in use » ; but this value in use is quite immaterial, because we have learnt, that money is money, exactly as far as it has an exchange value, apart from its possible value in use. We can now add that even its value in use is largely due to the fact of representing an official, or not official, money. This is the case e.g. of golden coins as ornaments. In these cases money comes first. It is true that a diamond is an important source of pleasure for the woman who wears it, and that the woman can be a relevant source of pleasure for the man who presented her with the diamond ; but the enormous price of a diamond is due mainly to its rare property of *keeping intact* a lot of richness during the time, i.e. of being money in sense (c). On the other hand, in archaic society the passage from ornament to money is very frequent. Among the ancient Germans the armillas were the most used form of money : some bracelets, which were also a sign of distinction. This passage is not difficult to understand. A badge, an ensign, a banner (i.e. a standard of value = *stendardum*) marks in general a *right to command*. Loosing its specificity, the same mark becomes either a simple ornament, or the generic commanding power of the money. In fact we say that a customer *commands*. The question is : why does he find so easily persons ready to obey ? Because these persons are anxious in turn to give orders, and trust that the symbol will be effective again.

14. VALUE DEPENDS ON THE FUTURE

This enables us to solve one of the most debated problems of economics, the meaning of the word *value*. The value of an object is expressed by the sum of a given currency you can obtain « realizing » it. Consequently, the general problem of value can be brought back to the problem of money value. Having settled that this is the value of an expectation, we must look for it *in the future*, and not in the past. Ricardo's theory of the value is well-known : the value of an object is equivalent to the quantity of labour needed to produce it. Marx took up this doctrine, with the obvious qualification that the labour must be socially useful. (In fact, one can toil along his life at a work, e.g. a novel, which results in no value at all). The right solution is simple : value is measured by labour, but not by the labour exerted for producing the object, but by the labour you can obtain selling the object out and spending the money you get thereby. The labour needed to pick up a gold nugget, you stumbled on by accident, is very small ; on the contrary, the value of the nugget is big, because a lot of labour can be *obtained* by it.

That is why the value of many commodities is so unpredictable. The labour needed to produce an object, being a matter of the past, can be accurately evaluated ; the labour you can obtain realizing it, depends on many circumstances, including a « propensity to work », which vary according to time, spot, health and habits. In this case, even big numbers cannot give a statistical certainty. If the trumpet of God should be heard, no economic value would stand up.

But let us focus our attention on a more proximate event : the reasons why a currency gains and keeps its value, or looses it during the time ; that is to say, why a practically useless object begins to be generally accepted as a security for a debt, whose debtor is yet unknown and whose content is undetermined. In the case of a *fiat money*, the origin is clear : the credit of the public power, entrusted with the collection of taxes. The question arises for a commodity money, whose usefulness as a commodity is anyway reduced, e.g. diamonds or shells. The origin of its monetary value can be a social one : the symbol can be used as a sign of the membership of a society, whose members furnish each other with services and presents, that must afterwards be rewarded. In some primitive societies the money grows up from a *present* which must be rewarded after a year with enormous interests. In other cases, somebody gives a pledge in order to enter a secret society, and the leaders can transmit it for some purposes to other members, eventually also outside the society : the sign becomes a « credit of the bearer ».

The procedure is easier, of course, if the pledge *was* already used as money. A more comprehensible origin of the money is a token given as a « security », for the restitution of a debt ; the token becomes the symbol of a « credit in general ». A peasant gives in summer his sheep to a shepherd, who brings it with many other animals in the mountains. A receipt is given to the peasant and is accepted by other operators, in exchange for something else. This is the origin of the banks. A goldsmith receives gold for work, and upon it guarantees bills for five times the value of the gold he owns, hoping that the creditors will not call for their property all at once. The receipts have become a currency.

15. MONEY AS « POTENTIAL LABOUR »

But we must consider a former conceptual step, by which *pecus* becomes *pecunia*. The traders accept the bills because they mean « gold », on the faith of the goldsmith ; but on which grounds do they accept *gold* itself as a sign of value ? After all, precious metals, gems, and so on, would not be precious at all, if they had not already a monetary capacity. Gems are money because they allow to postpone decisions, to escape the requisitions of Soviet Russia, to buy a hotel in New York and so on ; but why do they keep their value of inappropriate money (inside a society, of course, not on a desert island) ? Only because they are expected to be accepted again and again. This acceptance is the first origin of the value, and is the recognition of a debt. To the bearer of gold, gems, banknotes and so on, *something is due*. Gold is the sign of a debt, issued by someone who enjoys a big credit for himself, and gives out a part of it. But what is this debt ? And who is the last debtor ?

According to our banknotes, the issuing bank will pay to the bearer the corresponding value in gold. This statement is false, of course, but it expresses a relation between the issuer and the bearer, who are « joined together » by that symbol. The large credit of which the bank enjoys (or the State, or the rich gentlemen of the Isle of Man, who emitted money stamped on small pieces of leather) has been handed to you, for a very little part. But the real debt is not the gold, and the real debtor is not the issuer ; it is something much more indefinite, the « system », in which you expect that someone will do something for you. This is expressed sometimes by the saying that money is a « debt of nobody ». But the system fulfils its obligations only on certain conditions, the most important of which is that money should not be available at will, but only as the corresponding amount of a capacity of labour, which is limited by nature.

The coinage of rare metals, the choice of particuliar shells, and so on, are meant to guarantee the scarcity of these monetary symbols but not unconditionally. If Pizarro discovers exceedingly rich mines of silver in Peru, nobody will be able to prevent a rapid fall of the purchasing power of this metal. Apart from that, France in the XVIIIth century had a long tradition (going back at least as far as Philip the Fair) in *cheating*, favoured also by the adoption of a money of account, the *livre tournois,* to which no coin corresponded. Credits and debts were generally expressed in *livres*, but they were paid in *louis*, *écus* and so on. It was sufficient, therefore, for the public power to establish a different exchange rate between the *livre* and the *louis* to rouse monetary storms within an hour. One of the two most important purposes of the Scotsman John Law in introducing the banknotes was, on the contrary, to ensure a *perfectly steady money*, because the value of the banknotes was expressed in *livres* and could not be changed. (The second purpose was a Keynesian one : to do away with « idle money », §18) As you know, the result was the first (rather mild and nontheless scandalous) runaway inflation of modern times. At that time it was not quite clear what really insures to a currency its value ; neither is it completely clear today. In any case, this is not subject to the sure grip of power. Public powers can indirectly predispose some circumstances, but surely they do not dispose of the value of a currency by decrees.

The use of precious metals as species for the coins makes more difficult the abuse, but this is only a side of the problem. The πρώτον πσεύδος is the non acknowledgment of the *ideal character* of money, that induces to think of it as a quantity or a « mass » of some substance. The real quantity of a currency depends on its value, therefore you may not make the value depend on the quantity. This naturalistic error is common, both to the Keynesian and to the monetaristic school, although the fancy lucubrations of the first are by far more dangerous than the fussy calculation of the second, because, in accordance with the same error, the Keynesian school judges suitable to intervene, the monetaristic school judges suitable *not* to intervene.

Anyway, the value of money being founded on expectations, you cannot measure it as if it were a stock. Expectations are guided by something existing, but deal with something *not yet existing* ; then the value of money cannot be expressed by something that is there (as the number of units of a given denomination or even the set of all existing commodities). The failure of Keynesian and of monetaristic prescriptions in avoiding macroeconomical disasters proves that something is wrong in the doctrine.

But perhaps we can state just *one* reliable law about any currency : let us call it « the law of decreasing value ». This tendency can be more or less effective, can even be inverted in the short run, but it is absolutely sure. And it is proved by a *reductio ad absurdum*. If you had at your disposal a currency, whose value remains the same during a sufficient number of centuries, thanks to the compound interests you could assemble a credit that nobody could pay. A pound of silver lent at the beginning of the Christian age would return today a mass of silver bigger than the earth. An ironical will was written on this topic, in the XVIIIth century. This will, if executed, would have transformed within four centuries the whole population of France in *rentiers*. The conclusion is that, in the long run, debts are never solved ; and that is probably the reason why creditors are in a hurry to be paid. Have we found perhaps a real macroeconomical law ?

16. THE QUANTITY THEORY

Let us see what happens when money circulates. The debtor transfers to the acceptor a part of his credit by means of a symbol. This symbol in fact constitutes a *credit instrument*. The last statement may sound astonishing : when somebody receives a sum of money, is he not paid out ? What else can he require ? And yet, if he could not require anything by his money, his cash would be as worth as nothing. But this is not the case, because (as Adam Smith said), money *commands labour*. But what, if the *system* (that is the real debtor) does not obey ? If it does not furnish labour in the quantity determined by the previous value of the monetary sign ? When a single debtor does not fulfil his obligation, somebody looses a part of his wealth. The same if somebody lives above his means and consumes more potential labour than he can afford. In such cases a *well-defined* subject looses. But when the money itself, as a credit instrument, looses part of its worth, the whole system being liable for it, the system is the debtor and the creditor at the same time. It pays less than was expected, and you become poorer as a part of the system ; but with whom can you get angry ? The system does not owe money but labour, and if it is less solvent than before, you do not suffer a nominal loss ; and nevertheless you loose, together with all those who detained a share of that currency. *Inflation* is the appearance of that loss.

This is generally explained by the « quantity theory » : the same sum of money buys less, because the aggregate quantity of money has increased more than the quantity of available goods. The quantity theory compares the amount of a whole currency with the quantity of goods available at a given moment, as if they were two things. But, as we have seen, the money is not a scalar quantity, and inflation cannot be explained on this basis. We will explain it by a *lag*, which unavoidably occurs in the circulation of money. But for that purpose we must cast a glance more attentively on the quantity theory, in the classical form of Fisher's « equation of the exchanges ».

Irving Fisher wrote an identity (Marshall gives a different formula, with the same meaning) : MV=PT. M is the quantity of money, V is the so-called velocity of circulation, T is the amount of all transactions performed in a given time, and P the purchasing power of money. If you assume that V is constant (what of course it is not), P shall be inversely proportional to M. The quantity of *numerarius* grows, while the quantity of available goods remains the same : obviously the rate of exchange becomes less favourable. This is true, because we started from money as a commodity : but the *numerarius* has completely ceased to be a commodity, it is a number, and the situation becomes a little more complicated.

Fisher's formula is not an equation, but a definition. Were it an equation, it would be dimensionally correct – states Frits de Jong, in his excellent volume on *Dimensional Analysis for Economists* (Amsterdam, 1967) – and this had not always been the case. Schumpeter discovered in the papers of an economist of the end of the XVIIth century, J. Briscoe, a similar equation, which was dimensionally correct, *only* if you put V=1 ; in which case Briscoe's equation becomes identical with Fisher's one. Now a dimensionally incorrect equation is always inconsistent, but not all correct equations, from a dimensional point of view, are consistent. In fact a strong objection has been repeatedly moved to Irving Fisher : the purchasing power P is a mere *index number*, and numbers have no « dimension ». The same can be repeated for all pseudo-variables of Fisher's equation, which is therefore an identity. Frits de Jong writes : « It requires a thorough understanding of economic theory to decide whether we are going to assign dimensions to index numbers in a given context ». Well, I think that a thorough understanding of economic theory will prevent us from assigning a dimension to all pseudovariables of the pretended equation. The so-called « velocity of circulation » has little to do with a movement : it is a *frequency*, the *number of times* a currency is employed in transactions. Moreover, you may assume V as statistically constant only under very strong conditions. One of these is the stability of P, which is precisely the question !

What about the quantity of money ? It is relatively easy to count how many coined or printed pounds sterling are in circulation (M_0), but it is rather difficult to evaluate the number of pounds circulating under a different form : cheques (covered or uncovered), bills, drafts, bonds, and other forms of quasi-liquidity (M_1 , M_2, and so on). Fortunately this evaluation is immaterial : the number of pounds is no quantity at all, until you do not know how much they are worth. So it is a pure truism to say that the purchasing power depends on the quantity of money and viceversa.

The error of the quantity theory is to assume that M is a quantity of money of a given denomination, conceived as an independent variable. In fact, each time you buy anything in dollars, you evaluate the *purchasing power* of a dollar and *thereby* you emit an unconscious conjecture about the quantity of « really existing dollars », but you never could make the reverse. Even the Fed could not desume P from M.

Although non existing, money is active, but it has the power of an expectation, not the dynamical power of a « monetary mass ». The images founded on this analogy are totally misleading. Elementary textbooks furnish the students with a false picture of the situation, when they depict a *tub* in which the whole Gross National Income is poured, and the tub never gets full, because a « hoarding » process opens a tap on the bottom. Yes, we speak of liquidity, but money is not a mass of liquid in this sense. On the contrary, avoiding to conceive the money circulating in a given area as a bulk, that can be made smaller or bigger, you will probably obtain a more sensible notion of the forces exciting and depressing the economic activity.

17. A NEGATIVE PRODUCTION IS POSSIBLE

The event that aroused the fuss in economics was the big recession of 1929-33, when enormous labour forces and a big share of the invested capitals remained unemployed. Keynes discovered that this situation is consistent with a state of equilibrium, and that therefore we would vainly await that the economical forces do away with it, spontaneously. In fact, unemployed persons, by that time, accepted to work for reduced wages, but nevertheless the industries did not employ them, because they were not able to place their product on the market. People desiring a car were not able to buy it, because they had no job. Enterprises, ready to produce cars, did not produce anything at all, because they were not able to sell.

The situation is far from being absurd, although many explanations have failed in giving an account for it. Probably the production had to be *reconverted*, in order to reach a more satisfying equilibrium. Anyway, it was completely right that the equilibrium does not necessary involve full employment.

This fact unleashed unprecedented attacks against the « law of Say », and renewed a polemic against the thrift, a polemic that in fact was as old as the human world. J.-B. Say, at the beginning of the XIXth century, maintained that the production of an economic good always produces, at the same time, the monetary means for paying for it, considering that the price will always be consistent with the remuneration of the production factors. But after 1929 the prices were diminishing, but no sufficient amount of cash was put foward. Owing to the naturalistic fallacy we have spoken of, the conclusion was that a part of the remunerations was kept back from consumption (« hoarded » through the tap) and, of course, also from new investments, which would have given no return.

Later on, the law of Say, founded on an identity, was reinstated by some economists, but the whole question still requires an exhausting consideration, because many economical notions involved in it are founded on misunderstandings. The production does in fact create the currency needed to pay for it, provided you judge correctly *what* has been produced. A man who writes books (like me) can more easily answer this question : when the book sells well, it produces a considerable income ; but more frequently a big portion of the printing remains unsolved. What was really produced then ? Raw material, to be sent for pulping. In both cases the production has created the means needed to pay for it, but only for the product which was really produced, not for a bestseller.

The case is frequent also in other fields, but it escapes our attention when it concerns other types of necessities. If I produce bread, at a cost exceeding the price, I loose, of course, but I produce nonetheless something useful, that can be sold below cost. If I loose alone, I work out of charity, at my expenses, but from the point of view of the global economy this production must be labelled with *minus*, not with *plus*. Economists in general refuse to introduce in their matrices a negative production, because they argue : I can produce very little, or even nothing, but I cannot produce the contrary of something. This is a mistake : the mistake of confusing the physical production with the economical one, and furthermore of ignoring that both have a contrary : destruction. A physical production is a transformation, and each transformation is positive and negative at the same time : it produces order at the expenses of a bigger disorder (entropy). We judge useful this transformation when the disorder is carried away, otherwise we obtain a prevalent destruction. Economical productions are positive when they make richer the system (not some single trader), and physically they can also be a destruction (tomatoes are crushed, because by a lower quantity you obtain a higher global income). The meaning of the word « production » in physics and in economy is therefore analogous, but not identical, and in both cases it can be positive or negative. A macroeconomical production is positive if, and only if, the system becomes richer by it. This means that it becomes more disposed to expand the reciprocal collaboration of its components, *for the sake of convenience*. In other words the production is positive only if it increases the potential labour. In microeconomics, this corresponds to a growth of the capital value of the firm. Now, if I invest for producing some commodity the market does not demand, my expenditure is not an investment but a consumption, or, if you prefer, a waste, and the waste is a destruction. This can be seen in any production aiming just at assistance. Such an activity is not simply *less* remunerative than a production for the market, it is in fact negative, a real destruction, and its cost must be paid by somebody else who produces actual goods. The cost of this disguised destruction is sometimes so high, that it would be less expensive to pay the wages, leaving the workers idle at home. A pseudo-production can be accepted sometimes for ethical reasons, provided that everybody knows that this is not a device for producing less, but a measure for destroying more, and that this destruction must be compensated anyway.

18. NO MONEY IS IDLE

The cause of consumption has been pleaded also directly, without disguising it under the cover of a pretended investment. Arguing against the law of Say, Keynes' followers maintained that a part of the purchasing power created by the production is hoarded or concealed, instead of being turned to consumption and put again in the production circle. This thrift may seem advantageous to a single operator, but it is unfavourable on the whole, because the spared money gives a return only if it is invested ; but this will not be the case if the producers foresee that their production will remain unsold, owing to the lack of liquidity produced by the hoarding. This error is blamed as a « fallacy of composition » : what can be good for an individual, can prove calamitous for the community, because the single advantage cannot be generalized. Smith's « invisible hand » reversed, as you see.

This could be a brilliant explanation of the recessions, if money could really be stolen away by the hoarders, as if it were a physical thing. In fact, the basin on the hill cannot remain void because some ruffians have derived part of the water to their private tanks. Particularly the Austrian school has pointed out that money is never « idle », even if it is kept in a safe. I would like to add a sort of restatement of the law of Say : money is always as affluent as it is needed in any given situation. The case of a stock-exchange is the most evident. When a stock-broker complains that « money was scarce this morning », what is he really complaining of ? That he was not able to sell as many shares he hoped at the desired price. But, as for the shares effectively sold at the agreed price, the money was perfectly sufficient (and it needed not even to be present, but in its original form of credit).

In everyday transactions money is needed also materially, by lack of reciprocal knowledge and confidence. Then the absence of coins, banknotes, and so on, can actually produce some troubles, but if the buyer has really the credit he pretends to have, the difficulty is soon overcome. Very rich persons can go around the world without a credit card, if they are sufficiently known. Mark Twain's story *The banknote of a million pounds* shows that a miserable (but intelligent) beggar can live like a prince and get richer and richer just exhibiting a banknote of that enormous value (they were emitted sometimes by the Bank of England, for some special purposes), without being allowed to spend it. A department store having lost the challange to give the change for it, it was not even necessary for Mister Adams to show his title of credit. Was that an idle money ? Not at all.

Let us consider the case of commodities that are on sale, but are not sold, because the would-be consumers have no money to pay for them. Should we say, in this Keynesian case, that the *real* aggregate demand is insufficient, because somebody has concealed his money ? Capacity is unemployed, workers are unemployed, prices are diminishing and nevertheless commodities that are wanted remain unsold. This is a fact, and Lord Keynes was right in dealing with it, but it is not sensible to blame hoarding for it, because the aggregate real purchasing power of a currency in a system is always equivalent to the aggregate offer of goods and does not depend on the nominal quantity of circulating units. If this number falls (owing to the hoarding), either the real value or the so-called « velocity of circulation », or both, increase. And the hoarder can give his money to a bank (and in this case his money is not idle) or he can keep it in a safe. In this case he lends at no interest to the system as a whole, owing to the increase of purchasing power of the rest of that currency. If many persons are not able to work for buying what they want, it is not the quantity of money which is defective, but its employ, i.e. the productive organisation. The allocation of the existing resources is not the best, and the input of money that gives no return (as Keynes demands) can contribute to improve it, but not always and not without some precise qualifications. The cursed « hoarding » is probably, on the contrary, the beginning of the cure, because, in order to reallocate the resources a certain *delay* is needed : the hoarding fulfils this task both on the side of the buyers and of the producers. In fact, a correlation has been suspected between the periods of recession in Kondrat'ev cycles and the periods of « intense technological innovation » (according to van Dinjin, *The Long Way in Economic Life*, London 1938). Being given that the fundamental task of the money is to postpone decisions, the fact can easily be understood : the so-called hoarding is not the cause, but the effect of the recession, the cause being the necessity to wait for some innovation.

Nevertheless, the situation is quite disturbing : for the individuals, owing to the unemployment, which is humanly appalling ; for the firms, owing to the impending of chain-insolvencies. A Keynesian policy is therefore justified as a temporary device for preventing a vicious circle of insolvencies, but not (as many have said) to « determine » the general level of income.

19. INFLATION HASTENS ALL DECISIONS

Believing that the hoarders are guilty, because they imprisone a lot of currency (John Law compared this to a bleeding), the Keynesian scholars proposed to compensate this fault by an input of money with licence not to return home. Roughly speaking, they recommended to squander, producing goods or services that the market does not demand. Egyptian pyramids or the excavation of holes which are to be filled again are suitable. Is it a joke of the witty Keynes ? Not quite. The squandered money grants to the unemployed workers a credit towards the system, by which they become able to pay for goods that are available and really wanted. A credit towards the system is then generated by a « deficit spending » of the public powers and the system (not a single operator) pays for it by a diminution of the purchasing power of its currency. The expectation of such diminution obliges all operators to *hasten* their decisions, so that the whole economy is put in motion. Keynes says that no inflation necessarily occurs, if some unemployed workers and capacities begin to produce. Having rebuked the quantity theory, we cannot accept *in toto* this explanation. No doubt that, if some people are presented with the possibility to buy something, a « virtuous circle » begins (called « multiplier ») in which new workers produce goods to be exchanged with the added money, and so on. But the initial produce being not required by the market, nobody pays for it, the squandered money does not return, and it is the currency in this case that looses a part of its value. Nothing comes out of nothing. The Keynesian device will overcome a momentaneous stalemate, but it will never « create » a new income, as seem to think many authoritative theorists, apparently affected by a delirium of omnipotence. Even the effect of the multiplier is subject to quite different evaluations. According to the Brookings Institution, a billion of dollars added by the public administration means about three billions of additional NP within a year, and always more than two billions after four years. According to the Federal Reserve Bank of Saint Louis, the billion remains a billion for nearly nine months, but, after a year and a half, the NP will be a little *less* than if no billion had been spent. This enormous difference does not depend on mere mistakes, because the consequence of an outlay of money depends on *how* the money is employed. But my point is the following : if the money is spent for nothing, i.e. for a negative production, then it *can* produce a difference – sometimes also a positive one, if an acceleration is needed – but only through an inflationary expectation, not without it. Inflation (we will soon see why) produces substantial benefits to its starters (generally the public powers, and that explains the success of the Keynesian theories among the economists who aspire to a post of governamental counsellors) sometimes it can also be useful to aid some producers who otherwise would remain idle. But the hope of establishing a constant « trade-off » between inflation and full employment (i.e. : of paying the full employment by means of a planned inflation) expressed in the « curve of Phillips » (1958) has proved inconsistent by the so-called stagflation, after 1970.

20. « MONEY MATTERS » OWING TO A LAG

« Money matters » can have manifold meanings, the most of which are obvious. Inflation is the most important case in which the meaning of the motto « money matters » is not obvious. Within the horizon of a quantity theory, it means that the quantity of a currency is decisive. And, in fact, the life and death of many businessman and, what is more, the well-being of whole populations depends on it. But some qualifications are needed. Inflation is a rather recent word (less than hundred years old) and contains in itself the reference to a sort of *swelling*. The phenomenon has given rise to the quantity theory, centuries before, not only the word inflation was invented, but even the science referred to as political economy was born. Too much gold was mined, and all prices expressed in gold went up. Less pure silver was contained in a pound : all prices in pounds became higher. The nominal « quantity » of the currency grew faster than the goods available for being bought : it was inflation. With the invention of the fiat money the phenomenon became sometimes gigantic : the nominal quantity of a currency is expressed by a number, and arithmetics put no limitation to the bigness of a number. Without a logarithmic representation, the graphics referring to a hyper-inflation would compel us to go up as high as Himalaya. At the same time a curious (but not unaccountable) phenomenon occurs : the *real* quantity of an inflated currency – i.e. the quantity of goods you can buy with the whole amount circulating under that denomination – becomes more and more slight. In 1923, the issuing bank in Germany was not able to issue as many means of payment as it was needed for everyday transactions, although women were recruited to make into packages the banknotes. In other words, during a hyperinflation, money becomes *scarce*. Loosing its capacity of delaying decisions, the currency looses at the same time its end and its existence : it becomes vanishing, precisely because its existence is a potential and has nothing to do with the materials which symbolize it.

First of all, let us point out that the inflation is an economical fact, but its causes are certainly not. Inflation is a macroeconomical effect which does not depend on microeconomical transactions. It depends on a unilateral decision, like a robbery, and has exactly the same consequences of a counterfeit. The counterfeiter forges a « debt of nobody » for his own sake, paid by the currency. The public power makes the same when it « creates » money that will not return through the taxation. So the cause of inflation does not fall under the science of economics, but under the science of robbery, and its mechanism was perfectly explained before the birth of scientific economics. For instance, the Chancellor of the Kingdom of France, Henri-François d'Aguesseau, wrote in 1717-18, before the Law affair, his *Considérations sur les monnaies* (published in 1777), where both the causes and the effects of inflation were accurately described, principally on the example of Philip the Fair's conduct.

But the Treasury is not the only subject which enjoys the benefits of an inflation (as long as it does not become an hyperinflation). The real value of all credits diminishes proportionally and not all the creditors have an equal chance to cope with this diminution. Indexation was invented therefore, and gave some results in South America. Some credits can be corrected even *before* the price index ; but they can never be corrected by everybody in time. Even if the « monetary illusion » (according to which « a dollar is a dollar ») is disappointed, not all persons have the same legal and economic opportunity to provide timely. So the inflation has a « gradient », and the loss expands progressively through the economic system. If the money with no return were distributed in an instant and uniform way (« by helicopter », says L. von Mises), no economical consequence would issue, as well as when a heavy franc replaces 100 francs of before. But it is not so.

« Money matters », and some operators thrive and others commit suicide, because inflation is a *flux* of varying speed. The lag produces radical readjustments in all economic relations, also in the absence of any economical transaction. On the contrary, the macroeconomical phenomenon influences the microeconomic decisions, because everybody is obliged to get rid of his credits as soon as possible. Beyond a certain deadline, nobody is interested any more in an economic production, which would pledge a capital for a while. It is more convenient to fix a debt in a stock of goods that can be sold later : but this is possible only for lucky fellows, who contract debts thanks to their power. And the last result is the loss of any economic power also of the counterfeiters : no constraint to pay in advance the taxes will be sufficient to preserve its revenue, and, what is more, also the benefits of issuing money vanish, because they are proportional to the *real* purchasing power of the whole stock of the currency concerned, and this, as we have seen, in hyperinflations rapidly decreases. The economic policy of all States inclines therefore to plan a low rate of inflation but it does not always succeed.

CONCLUSIONS

I

Economy is essential in human life, because it consists in planning one's behaviour instead of reacting immediately to the facts. Many other animals, besides mankind, provide to their future accumulating consumption goods or capitalizing their work in nests ant-hills and so on. The behaviour of a wasp that stings a spider in its nervous ganglions without killing it, in order to lay its eggs in a living but motionless body, that shall feed the larvas, is very near to the foresight of a family which opens an account to be used when the sons attend University. But what is peculiar to the human kind is the inventiveness in providing for the future, the freedom in the choice of the means, and particularly the reciprocal agreement and organization of these means « for the sake of convenience », which enormously increases the cooperation with other people.

The possibility that different opportunities meet in the working organization is provided by the civil law and protected by the criminal one, so the economic behaviour is strictly conditioned by law in general. Law should therefore allow a freedom of negotiation as large as the public order consents. The limitations, albeit frequent, should always rest on the general interest of the population (e.g., defence against pollution, and other relevant « externalities »). Provided that such nuisances are avoided, the freedom of trade is the most profitable way of supplying human kind with all the relevant commodities and services. At the same time, it is an educational way to « virtue », as it was said in many textbooks of the past, whose expression today is perhaps apt to arouse laughing (let us remember *Thrift*, by Samuel Smiles, 1875), but whose content is very sensible in pointing out the significance of economy for the realization of freedom.

The antichrematistic attitude of many authors (encouraged in the last two centuries by a socialist propaganda, which in fact recommends different ways of getting rich and influent, which are really worse) is justified as far as it points out that there are values more important than the economic one. This is indisputable and should be taken as a matter of course. Economy is meant for the acquisition of means, not for the determination of ends, and no means could have a value, if there were not some ways of living which have a value in themselves. Even the behaviour of very rich people, who persist to pursue a wealth from which they cannot receive any additional enjoyment, confirm this truth : they act according to the « theory of games », and what they aim at is not the value of the money, but the value of a game.

Another inconvenience of the accumulation of wealth is the enormous power of evil it entails (as well as of good). More education is therefore particularly needed for rich people, because they enjoy a larger freedom. The habit of treating richness as a « means », and, not as an end, should be developed among them with a particular care, because the affluence itself could make this truth more obvious and therefore unperceived. Both the greed for the greed's sake and the propensity to conceive life as a perpetual game of « Monopoly », must be fighted against, and the best way to do it is to distinguish between the stage in which wealth is produced and the stage in which it is enjoyed. In the former, profit is not only permitted, but duty-bound. In the latter, liberality and unselfishness should be the chief characters. The morality of American self-made men of the pionneers' age is always valid. It is of course a rather abstract distinction, and we can wonder e. g. how far a competitor may push a rival on the road of suicide, even by lawful acts. But in any case, to confuse the economic activity with public charity is the best way to prevent the former from supplying the means for the latter. A consequent conceptual error is to believe that economy has to cope with a « just » distribution of wealth. Economy can distribute the income only automatically, according to circumstances that have nothing to do with ethics (as von Hayek repeatedly pointed out). Moreover, the distributive justice can be exerted only by somebody who is the legal owner of the good (in the case of a State, the ownership is generally got by taxation), by a unilateral decision of him : this ought to be *just*. This decision is essentially different from an economic transaction. There is therefore no particular « business ethics », but only a general ethics whose rules apply all to business.

This ethics commands also to obey to all the laws in force, but the fact that these laws are sometimes senseless, and not generally followed, puts some questions difficult to be solved by general rules.

A more serious question is the following. Economics deals with the results of free agreements, not with enforcements of any sort. But the economic operators do not enjoy freedom all at the same degree. The richer one is, the more he can wait : he enjoys therefore more freedom. No decision inspired by an immediate constraint, like hunger, can be said to be free. In this sense the economical order is unjust, or, better unequal. This is the strongest argument in favour of a welfare state, where everybody is endowed with a certain degree of economic independence. The difficulty is that this independence cannot be conceived and implemented independently of the conduct of the person concerned, otherwise all inducement to produce would stop.

The economic system can be compared to a vector-field, in which strong differences of potential are present. This difference has been recognized as necessary to the economic activity, and it is therefore accepted also by the followers of an equalitarian ethics. The tension created by the difference of potential goes up and down in periods of boom and of recession. In the former, everybody is more inclined to agree with other people to any sort of transactions, the contrary occurs in the latter. This is the key for understanding the economical conduct, which has always to do with vector-quantities and potential states, not with mere scalar quantities.

For instance, the computation of NGP in different countries (invented in order to make possible the allottment of contributions to international organizations) is highly deceptive, being based on scalar quantities, which give not the same result everywhere. In different systems the tension can be very different at the same moment : very high in USA and very low in India, owing to the strong negative *social entropy* of the last system and to the strong *economic* negative entropy of the first. The gap is filled slowly by the interaction of different systems ; and it can be filled more rapidly if in a given country the social negentropy yields suddenly in favour of an economic one. This happened, e.g., in Japan in the last century, and this is happening now in some other countries of the far East.

Many objections against the market economy – i.e. against the economical activity as such – prove right, and the remark that market cannot work out every problem is quite obvious. But we hardly get aware of the enormous quantity of problems it does resolve, because it does it quietly. And above all we should always remember that there is just one different option to the economical organization : this option is slavery, no matter how it is called.

II

Economy, as it was strictly defined here, is peculiar to the human kind, and it is spread among all human societies. Everywhere are present some tokens which symbolize : « I have a right to be served by you ». This symbol becomes money, in strict sense, when neither the person of the servant nor the nature of the service to be performed are determined, until two or more persons agree about it and about the value of what has to be done. Money can consist of a large variety of symbols (it can possibly remain also implicit), because money is a number ; and « what has to be done » implies the picture of a future which must become a present. For that purpose a *lag* must be overcome, and money fulfils this task. All living beings take steps concerning their future, but they do it instinctively, they do not project (= « cast before them ») their actions. Men do this and moreover in a peculiar kind, by way of (relatively) free transactions. The necessity of overcoming a lapse of time in order to reach one's ends is typical of a finite creature. We must *wait*, because we are not omnipotent, and the only persons who have not to wait, when contriving their plans, are the three persons of the Christian Trinity. That's why they do not want money as means ; and the utopians, who dream to do away with money also in human relations, are not completely aware of this difference.

We must wait, otherwise we could not even exist : we should immediately reach the end, *viz* the death. Well, money is a *means* (= something that « lies between ») for making busy our waiting. A means symmetrical to space, because also a spacial distance divides us from our aims. We reach our ends projecting our action in space, but any action requires time. What overcomes the spacial distance are physical means, what does the same with respect to the time is called richness : an ideal device which the common language identifies with the « means » *par excellence*.

« Time is money » is a popular slogan, not only among businessmen. The statement can and must be reversed : money is time. It is the concretion of time. It makes concrete all our projects, allowing their implementation. It transfers our richness in time, by the payment of an « interest » (i.e. something that « exists between »). Money takes time to be gained, but it makes our time « free », when it is spent. We are not the masters of our time, we can nevertheless dispose of it, if we have « the means ». Thereby our finitude becomes a power, also a power of a very peculiar sort, a social and potential power. There is a tension, that binds people together, everybody aiming at his own ends, but all striving together towards a result that nobody had in mind at the beginning. This movement engenders the human civilization.

BIBLIOGRAPHY

DE JONG F., *Dimensional Analysis for Economists*, Amsterdam 1967.

EINZIG P., *Primitive Money*, Oxford 1996^{2}.

FAURE E., *La banqueroute de Law*, Paris 1977.

FOSTER T.S., *From Savagery to Commerce*, Oxford 1930.

FRIEDMANN M., *The Quantity Theory : a Restatement*, Chicago 1956.

GASPARINI I., *Il sistema economico*, Roma 1978.

GEORGESCU ROEGEN, *The Entropy Law and the Economic Process*, Cambridge (Mass.), 1971.

HAYEK (VON -) F.A., *The Road to Serfdom*, London 1944.

- *Individualism and Economic Order*, Chicago 1972 (with bibliographical references).

HICKS J.R., *Value and Capital*, Oxford 1939.

KEYNES J.M., *The General Theory of Employment, Interest and Money*, London 1936.

KNIGHT F.H., *On the History and Method of Economics*, Chicago 1956^{2}.

KNOPF K.A., *A Lexicon of Economics*, San Diego 1991.

LITTLE D. (ed.), *On the Reliability of Economic Models*, Essays in the Philosophy of Economics, Dordrecht 1995.

MATHIEU V., *Filosofia del denaro dopo il tramonto di Keynes*, Roma 1985.

MISES (VON-) L., *Human Action*, New Haven 1949^{3}.

PATINKIN D., *Studies in Monetary Economics*, New York 1972.

PHELPS E.S., *Seven Schools of Macroeconomic Thought*, Oxford 1990.

REDMAN D.A., *Economics and the Philosophy of Science*, Oxford 1991.

RICOSSA S., *Teoria unificata del valore economico*, Torino 1981.

ROUSSEUS S. (ed.), *Inflation : its Causes, Consequence and Control*, Wilton (Conn.) 1968.

SAMUELSON P.A., *Economics*. An Introductory Analysis, New York 1948 ; 14th ed., with W.D. Nordhaus, 1992.
SEDILLOT R., *Histoire morale et immorale de la monnaie*, Paris 1989.
SIMMEL G., *Philosophie des Geldes*, Leipzig 1912.
SRAFFA P., *Production of Commodities by means of Commodities*, Cambridge 1960.

La Technique d'aujourd'hui et les problèmes d'esthétique

Tomonobu Imamichi
Centre international d'études comparées de philosophie et d'esthétique, Tokyo

Introduction

Comme en général dans la philosophie, les problèmes de l'esthétique sont permanents, parce que l'esthétique est au fond une des disciplines de la philosophie, et la philosophie traite de problèmes essentiels qui doivent être pensés constamment, sans cesse, à travers tous les temps. Aux mêmes problèmes classiques, nous pouvons naturellement chercher des réponses différentes de celles qui furent proposées dans le passé. Pour la nouvelle recherche sur l' esthétique moderne, nous pouvons présenter nos nouvelles études sur les problèmes traditionnels comme le beau, le sublime, les arts et le jugement esthétique, etc. Mais une telle recherche ne peut pas être universellement intéressante pour tous les esthéticiens, parce que l'impact du problème ou le pathos provoqué par un problème est individuellement différent chez chacun des esthéticiens.

Cependant, il y a un problème qui suscite l'intérêt universel. C'est le problème de la technologie moderne dans le domaine de l'art. Un homme qui vit dans la société d'aujourd'hui, s'il a de l'intérêt pour l'esthétique, doit, à notre avis, nécessairement penser à la relation entre les techniques d'aujourd'hui et le problème de l'esthétique. Ceci explique le titre de notre recherche.

La sphère de notre recherche

À lire notre titre « la technique d'aujourd'hui et les problèmes d'esthétique », on pourrait penser qu'il s'agit d'un art mécanique,

P. Kemp (ed.), World and Worldhood, 149-168.

technologique ou de la cognition quantitative de la perception esthétique. On pourrait aussi penser que notre étude est limitée à un cadre analytique de la cognition scientifique face à l'expérience esthétique en tant qu'elle est comprise comme expérience sensorielle. Aussi, nous rappellerons ce que nous avons déjà dit : l'esthétique est, au fond, une des disciplines de la philosophie. L'esthétique est métaphysique parce que la philosophie est métaphysique. Comme nous l'avons déjà expliqué dans nos autres écrits, la métatechnique est nécessairement une nouvelle discipline philosophique dans l'environnement technique de notre société moderne. Pour développer notre essai, notre pensée est nécessairement métaphysique ou métatechnique[1], en ce qui concerne la dynamique de l'expérience esthétique. Pourquoi ? Parce que l'expérience esthétique est une perspective profonde qui commence avec la perception sensitive comme objet de la science naturelle et qui se développe par configuration sémantique comme objet d'analyse. Elle s'élève comme valeur artistique en tant qu'objet d'interprétation et s'intègre à l'idée du beau comme corrélative à la pensée métaphysique ou métatechnique, la pensée calonologique.

Dans notre essai, il ne s'agit pas seulement de la métamorphose de l'art technologiquement forcé, mais aussi des problèmes essentiellement philosophiques ; par exemple, la temporalité qui est essentielle pour l'expérience esthétique, la comparaison axiologique entre la fonction technologique et la valeur transcendantale de l'art, la pensée existentielle du concept *kairos* et la nouvelle tâche de notre société contemporaine, celle de l'esthétisme. Ainsi, à mon avis, la sphère de notre essai est tout à fait philosophique, réflexion nécessaire pour la culture de notre société d'aujourd'hui.

---

[1] À notre avis, quatre nouvelles disciplines philosophiques sont nécessaires : 1) la métatechnica, 2) l'urbanica, comme philosophie de la cité, 3) l'éco-ethica, comme éthique philosophique pour le nouvel écoumène humain dans l'espace cosmique techniquement élargi, 4) la calonologie, exposée dans la note 3.

La metatechnica est une science philosophique visant une réflexion cognitive et une projection inventive de la technique comme technologie ou technocratique ; on peut peut-être dire aussi à la place de metatechnica, metaphysica de la technique en général. Mais à mon avis, le terme « metatechnica » est meilleur, parce qu'il désigne la méta-pensée de notre nouvel environnement technologique, justement comme la métaphysique est dans son fond la méta-pensée de l'environnement comme *physis*, à savoir la nature. La metatechnica doit alors être méthodiquement différente de la métaphysique.

TERMINOLOGIE

Qu'est-ce que l'on entend ici par les mots « technique » et « esthétique » ? La technique n'est pas la technique dans l'art, ni l'habileté de l'art, ni la technique au service de l'art, mais la technologie, dont le spécialiste se nomme ingénieur. L'esthétique ici n'est pas l'esthésique[2], mais la calonologie[3], qui est au fond la métaphysique du beau et qui comprend, comme l'une de ses parties, la philosophie de l'art, dont les propédeutiques auxiliaires sont les sciences des arts, par exemple, la musicologie, l'histoire des beaux-arts, la science poétique, l'architecture, etc. Il ne s'agit donc pas de la technique artistique dans le cadre de la science des arts, mais de l'influence réciproque de la technique industrielle moderne et de ce qui appartient à l'esthétique calonologique non moins qu'artiologique.

Cette réflexion est pour nous, aujourd'hui, non seulement intéressante, non seulement importante, mais nécessaire, parce que la connaissance claire des causes de la transformation remarquable de l'art moderne de notre temps et de la société technique, ainsi que la compréhension juste des significations positives et négatives de cette transformation, peuvent nous donner au moins une indication pour orienter la pensée vers une vraie réflexion sur la beauté et sur l'art humain. En effet, on écrit beaucoup sur l'art, mais c'est presque toujours seulement sur des phénomènes périphériques de l'art. On ne sait pas ce qu'est l'art. Puisqu'on ne connaît pas l'essence de l'art, on oublie la nécessité de la connaissance de la beauté qui n'est pas le plaisir sensible. Nous ne pensons pas que la beauté soit réalisée seulement dans l'art, et nous ne pensons pas que le but de l'art soit seulement la beauté. Mais en tout cas l'art est un moyen concret pour chercher la beauté. Et la réflexion qui veut aller de l'influence de la technique sur les phénomènes

---

[2] Le mot « esthésique » au sens de l'esthétique sensorielle est utilisé par P. Valéry et R. Bayer, *Traité d'esthétique*, Paris, Colin, 1961, p. 38.

[3] La calonologie, c'est la philosophie du beau ou la métaphysique de la beauté. J'évite le mot *Kallistik*, parce qu'il suggère la nuance d'une science de la norme comme l'éthique ou la logique. Car la calonologie n'a pas seulement affaire à la recherche de la norme, mais aussi à la méditation ontologique de l'être de la beauté. *Cf.* nos articles dans les *Actes des Congrès internationaux d'Esthétique* en 1956 à Venise et en 1960 à Athènes et « Der logische Grund der Schönheit » dans la revue *Studies in Philosophy*, Fukuoka, 1962. Cet article modifié est le Ve chapitre de mon livre *Studia comparata de aesthetica*, Tokyo, 1976.

artistiques à la fonction de l'art lui-même doit nous aider à penser la beauté.

AUTO-INSTALLATION[4]

La technique regardée en relation avec la révolution industrielle du XVIIIe siècle n'est qu'un élargissement de l'instrument humain. Elle renouvelle le don de Prométhée ! Elle offre une arme avec laquelle l'homme dispute le terrain aux autres animaux dans le monde de la nature. Mais ce sens de la technique est déjà dépassé. La technique moderne est autre chose qu'une extension de l'équipement humain, elle est une auto-installation d'un système cohérent du machinisme, elle produit un monde dont l'influence sur l'intériorité humaine est si grande qu'à l'examiner on peut éclairer les problèmes de l'esthétique actuelle.

LE RENVERSEMENT DU SYLLOGISME PRATIQUE

Sans doute l'intérieur humain reste un abîme même quand une machine nous montre ses ondes cérébrales. Mais l'intérieur humain si mystérieux ne se cache pas toujours. Il se montre par la forme de l'acte où il s'épanouit. De sorte que l'acte est une solution pratique de la pensée morale qui suppose nécessairement le syllogisme pratique, dont la formule classique se trouve dans l'*Éthique à Nicomaque*[5]. A est désirable. Or, p, q, r, s, et t réaliseront ce A désiré. Donc je choisis p pour quelque raison, comme le moyen pour arriver au A désiré. Dans ce syllogisme pratique la mineure est l'horizon de la liberté du choix, dont l'objet est un moyen à exercer pour la fin. En d'autres termes, la fin comme idéal a la primauté sur les moyens effectifs entre lesquels on

---

[4] Le mot auto-installation que nous avons proposé au colloque international d'esthétique (Dufrenne, Imamichi, Vattimo, Mingue, Revault-d'Allones. Tertullian, Per Hamberg) à Amersfoot près d'Amsterdam, est utilisé ici pour souligner le caractère autarcique et autonomique de l'organisation technologique. Nous ne sommes pas sûrs que cela soit un néologisme. En tout cas, ici, ce mot auto-installation veut dire la cohésion technologique fermée et autonome comme société contemporaine.

[5] La liberté du choix chez Aristote ne concerne pas le but de l'action, mais le moyen. L'objet du choix porte sur la réalisation de ce que veut la *boulésis.* Cf. Aristote, *Éthique à Nicomaque,* 1111 b 26-1112 b 17, et aussi notre article : « Die Notizen von der Metamorphose der Klassischen Ethik bei den griechischen Kirchenvätern », *Studia Partistica,* V, Berlin, 1962, et mon livre : *Betrachtungen über das Eine,* Tokyo, 1968.

choisit. Cette primauté rend possible un progrès rapide de la technique comme moyen. Quant à ce progrès de l'effectivité du moyen, il y a justement un saut qualitatif de l'*instrument* machinal au *monde* machinal, ou, autrement dit, il y a un saut génétique de l'outil hétéronomique à la force autonomique. Pour cette raison, la primauté de la puissance sur le but, c'est-à-dire la primauté du moyen sur la fin, apparaît comme un phénomène moderne dans le domaine de tous les actes humains. La structure du syllogisme classique de la pensée pratique se heurte à une grande difficulté. Le genre humain d'aujourd'hui possède une force systématique puissante (l'énergie atomique ou électromécanique et le capital de l'économie mondiale, etc.). Une inversion[6] d'ordre des prémisses dans le syllogisme pratique s'opère ainsi : nous avons un moyen P. Or, P peut réaliser a, b, c, d et e. Donc je choisis a, pour quelque raison, comme la fin de ce moyen P acquis. Cette fois aussi, la mineure est l'horizon de la liberté du choix, dont l'objet cependant est une fin à réaliser. Mais la fin n'est plus un idéal transcendant qui est désiré, elle n'est qu'un des aboutissements naturels d'une puissance physique. Le moyen détermine alors une fin à réaliser. Il prend le pas sur la fin.

Cette souveraineté de la force exerce une influence considérable sur des problèmes esthétiques modernes, spécialement sur l'attitude créatrice dans l'art, car la création artistique est aussi une des formes de l'acte humain. La fin de la création aujourd'hui n'est pas toujours la poursuite d'un idéal transcendant comme principe formateur, mais très souvent ce n'est qu'une réalisation arbitraire d'une possibilité déjà acquise. Ce n'est plus un mouvement spirituel vers l'infinité mais seulement un jeu limité par une chose finie. Alors, il ne s'agit pas de valeur verticale, mais de nouveauté horizontale. On tient donc toujours l'originalité en haute estime, même quand elle serait simplement singulière. Or, dans l'art, on devrait au fond estimer non pas la singularité excentrique mais la perfection « incentrique » verticale. Mais on confond plutôt l'unique avec l'individualité, la perfection avec l'originalité. Ce phénomène est déjà un symptôme pathologique où la création et la production, l'atelier et le laboratoire, le salon d'exposition et le compte rendu d'exploration sont mis sur le même plan. Et pourtant,

---

[6] Nous avons déjà discuté en détail de l'inversion de la structure logique pratique en 1954 dans la revue *Tetsugaku zassi* (revue philosophique japonaise) t. 68, n° 718, pp. 1-19 ; cf. notre livre : *L'identité et son auto-formation* (en japonais), Tokyo, 1970.

si l'art est aventure, ce doit être pour l'essor libre vers le transcendant, et jamais pour l'accentuation de la simple différence horizontale de l'individu. Mais pourquoi le mot transcendant est-il utilisé ici ? Par lui, nous entendons seulement la valeur comme principe formateur, introuvable dans le monde de la chose et dans le domaine de la subjectivité, mais appartenant au domaine de l'objectivité, sans pourtant être le monde de la chose ; cette objectivité peut se nommer un transcendant.

La structure logique de l'acte humain, dont l'orientation est, par la technique moderne, ainsi déviée du choix d'un moyen vers l'énumération sélective des fins possibles, remplace l'idée du beau comme principe formateur par l'auto-revendication, l'auto-insistance du moi, car celui qui a la force et les fins possibles, c'est le Moi appuyé par la technique moderne.

D'où deux résultats esthétiques. D'une part, l'expression[7] du Moi, qui s'oppose à la représentation stricte par la technique mécanique du monde extérieur, est devenue le seul terme authentique de l'art moderne. Par cette idée d'expression, l'art a pu approfondir sa fonction jusqu'à la possibilité d'exprimer la disposition de l'intérieur humain. Ainsi l'art se distingue-t-il de la technique, qui peut bien décrire des mouvements psychologiques, mais qui ne peut aucunement décrire l'état d'âme. Vraiment, c'est un progrès positif auquel l'art est conduit par la technique moderne. L'art en appelle à l'existence et devient une quasi-religion. L'esprit qui fuit le tracas technique et la crise politique désespérée veut se plonger dans la fascination existentielle des images divinatoires de l'expression, afin d'oublier sa vie bruyante au grand jour. L'art devient un jeu de la nuit privée pour penser au rien. Il libère un moment le prisonnier de l'angoisse, presque comme de l'opium. Mais en devenant si égocentrique, il a perdu de plus en plus son intentionnalité transcendante. Si l'expression est simplement l'exhibition artistique d'une subjectivité limitée, elle n'est qu'une représentation immanente du moi humain : son intentionnalité n'est plus transcendante.

D'autre part, l'œuvre d'art, qui était le moyen par lequel on figure son idéal comme sa fin, est maintenant devenue sa propre fin. L'idée de « l'art pour l'art » est ainsi une des conséquences de l'inversion entre les

---

[7] Cf. notre article : « L'expression et son fondement logique » dans *Revue internationale de Philosophie,* Bruxelles, 1962, n° 59, fasc. 1. C'est le IVe chapitre de notre livre *Studia comparata de aesthetica*, 1976.

prémisses pratiques de l'acte humain. Comme idée, l'art pour l'art est à mon avis naturellement supérieur à l'art pour l'idéologie politique ou religieuse qui force l'art à servir l'autorité mondaine au lieu de figurer le vrai transcendant. Mais l'art pour l'art, cela veut dire aujourd'hui que la technique machiniste fixe la portée de l'art. Et l'autisme de l'art isole alors l'artiste. L'art et l'artiste, non moins que la beauté, révèlent pour la société de nos jours sa surabondance et son luxe. Quelle erreur, et quel dommage pour l'humanité ! Car l'art est vraiment nécessaire pour la vie humaine : on n'a jamais vu une société sans art même dans les temps les plus primitifs. Et pourquoi ? Cherchons-le maintenant.

Quand on pense non à l'art en soi, mais aux alentours de l'art, cette primauté du moyen est, semble-t-il, très favorable pour l'art. Cette puissance de la technique, en sécularisant son but, révolutionne la situation du monde quotidien. Croissance constante du niveau de vie, élargissement des loisirs... L'art est partout sollicité. De plus, la technique qui récuse la primauté de la fin en général a libéré les œuvres d'art de leurs buts originaires et originaux. Dans le dernier siècle on a inventé le musée où l'on expose objectivement beaucoup d'œuvres d'art, arrachées à leur patrie et à leurs fonctions. Et le musée s'ouvre : grâce aux techniques de transport, on peut voir et écouter toutes les œuvres d'art du monde entier : l'étendue de notre champ de vision est plus vaste que ne l'était jadis celui des maîtres. On connaît concrètement toutes les tendances artistiques. N'est-ce pas un bienfait ? En partie oui, tant qu'il s'agit des alentours de l'art. Mais en même temps ce groupe systématique d'esclaves mécaniques est un groupe organisé de saboteurs qui attaque le cœur de l'art. Et comment ?

LES DEUX VISIONS DU MONDE

L'existence du Moi, le thème fondamental de l'expression, est entendue sous forme d'angoisse, d'inquiétude ou d'espérance, comme une modalité de la conscience humaine. L'essence de la conscience humaine est la temporalité. Or, la technique moderne accélère le temps si fortement qu'elle anéantit la plénitude de l'instant, et compromet par là la conscience humaine. L'homme doit agir suivant la vitesse du travail machinal à la chaîne, et sa réaction rapide au mouvement mécanique lui est plus nécessaire que sa pensée calme. Ainsi la pensée dont la profondeur mesure la dignité humaine est identifiée à la lourdeur. La méditation est l'hésitation. L'homme est considéré comme une fonction physique sans pensée, et non pas comme une personne. L'homme et la machine s'identifient du seul fait qu'ils sont tous les deux une organisation fonctionnelle qui engendre un certain effet statistiquement et stochastiquement prévu. Cette identification de l'homme et de la machine est, au lieu de la classification traditionnelle où l'homme et l'animal s'égalent par le concept d'organisme, devenue le principe de la classification totalement neuve par le concept de fonctionnalisme. D'où deux visions du monde (*zwei Weltanschauungen*) : a) la perspective machinale sans ombre de l'animalité, l'horizon mécano-mathématique sans liberté psychique, et b) son antagoniste extrême, c'est-à-dire l'extension de l'inconscience animale, l'espace de la subconscience. Dans les deux cas, il n'y a pas de place pour l'humanité. L'art abstrait et l'art non-figuratif (sans circonscription rationnelle) sont un reflet de ces *Weltanschauungen* dans le domaine de l'art visuel. Car la statue sans visage est aussi un signe de la perte de la personnalité. Une telle tendance de l'art et du goût d'aujourd'hui doit s'interpréter non seulement par le développement propre de la forme ou de la manière artistique, mais plus profondément par cette révolution de la *Weltanschauung*.

Dans la perspective mécaniste, on considère toutes les choses selon l'ordre méthodiquement bien réglé. Donc, dans tous les genres de l'art, le positivisme méthodique et le *pathos* de la découverte de la méthode sont à la mode. Par là, le niveau universel de la connaissance de l'art ou de l'habileté dans chaque art s'est certainement relevé. Mais le coup de maître, une vraie œuvre d'art, est une cristallisation verticale qui coupe la continuité horizontale de l'histoire. L'éminent est toujours hors des limites du cadre de la méthode universelle. Chez l'esprit créateur d'aujourd'hui, alors, l'impérialisme de la méthode suscite le désespoir. Comme résistance à ce méthodisme, une atmosphère de libéralisme dévoyé règne autour des artistes. Ironiquement on doit dire que la technicisation méthodique du monde amène la détechnicisation de l'art.

Dans l'horizon opposé de l'inconscience animale, il n'y a pas l'unité du monde, mais seulement une circonstance fragmentaire : une angoisse indéfinie, une joie instinctive. Et comme on ne retrouve pas ici le déterminisme cruel du calcul mécanique sans liberté, cet horizon est devenu le lieu où l'on fuit le monde technique. On cherche à s'évader afin de s'oublier au lieu de se libérer. La nostalgie de la liberté se plaît à affirmer l'indétermination propre au domaine de l'inconscience, bien qu'elle ne puisse s'en satisfaire, parce que la liberté n'est pas égale à l'indétermination. Beaucoup d'œuvres musicales de l'atonalité et beaucoup de peintures du surréalisme sont plus ou moins une expression représentative de l'indétermination vague de cette scène. Retour aux mouvements psychiques du préhomme, irrésolution psychologique. Ici le dessein, qui ordonne strictement des éléments, n'est pas nécessaire. Nécessaire est un air indéfini qui n'est qu'un état désordonné. Ironiquement on doit dire encore que la technicisation mécaniquement ordonnée du monde a désordonné la scène de l'art.

Car l'homme dans le système technologique n'est qu'une fonction machinale, il n'est plus un être inéchangeable comme la personne, mais il est déjà, comme pièce de rechange, un élément échangeable. Il n'a plus de mission, plus de profession, mais seulement une position sociale qui n'est qu'un point dans la circulation automatique. C'est le règne de l'anonymat et du nombre. Autrement dit, c'est le *pathos* de la répétition échangeable, l'approbation de la réitération, la louange de l'immortalité de la fonction par la magie technique. Cette magie de l'immortalité fonctionnelle requiert de grands sacrifices. On altère la pureté du chef-d'œuvre en le reproduisant mécaniquement. On abolit l'extase de sa rencontre unique. La reproduction, c'est le nivellement de l'expérience extatique ; c'est le monde pour la précision, mais ce n'est jamais l'horizon pour la vérité. Par la reproduction, tout le monde peut bien éprouver partout une exaltation limitée, mais seule une expérience directe de l'original donne une chance à l'enthousiasme authentique. On peut s'oublier même auprès d'une copie. Mais on ne doit pas confondre l'oubli de soi et l'exaltation de soi[8].

LA TÂCHE DE L'ESTHÉTIQUE

L'homme, dans le système technocratique, perd la pensée. Mais l'homme sans pensée, que fait-il ? Il agit justement comme une machine sans aucune immobilisation temporelle, il prend le chemin le plus court vers son but ; il le choisit instantanément en négligeant les autres chemins qui ne sont pas si efficaces. Ici se trouve un mode tout nouveau d'abstraction. Non plus l'abstraction logique, mais l'abstraction technique ; on

---

[8] Nous avons discuté de l'inauthenticité de l'art aléatoire avec M.D. Morgan de l'Université du Texas qui nous a répondu dans le colloque sur la situation de l'art moderne sous la présidence de J. Gantner au Ve Congrès international de 1964 à Amsterdam. La contribution de V.M. Ames est aussi intéressante. On ne peut pas oublier de mentionner le livre très judicieux de Ch.-P. Bru, *Esthétique de l'abstraction*, Paris, 1955.

n'abstrait pas l'universel du particulier, mais on abstrait l'efficacité en abolissant les autres valeurs, et au terme de cette abstraction, on ne garde que le résultat en abolissant tous les labeurs du processus. La montée du funiculaire est un symbole de cette abstraction technique. Certes, c'est très utile, et même l'enfant invalide peut y trouver la joie d'une ascension ; mais en même temps on perd le sens de l'effort personnel de l'alpiniste. Cette abstraction technique domine notre temps. Et son influence sur le domaine esthétique est terrible. Car elle ne tient aucun compte de la valeur du processus laborieux de la création artistique, où l'on essaie de cristalliser son œuvre par sa liberté et sa recherche. D'où une production aléatoire, la liberté s'en remettant au hasard de la matière, comme dans la musique stochastique ou le « dripping ». Ce n'est qu'un cri, ce n'est pas un chant, ce n'est pas un caprice, ce n'est qu'une bêtise. Le résultat peut être joli, et même beau, mais ce n'est pas de la musique, ni de la peinture, ce n'est qu'un jeu des tons ou des couleurs. Un chat peut produire un effet joli sur les touches du piano, quand il y joue ; peut-on dire sérieusement que c'est une musique relevant de l'art ? Placer délibérément le chat sur le piano n'est pas faire de la musique. Un tel projet que des artistes ultra-modernes expérimentent fièrement n'est à mon avis qu'une préparation matérielle de la scène où la beauté naturelle peut apparaître par hasard. Ce n'est qu'une collaboration de l'homme à la beauté de la nature : ce que le jardinier fait pour la fleur.

Mais d'autre part, on ne doit pas oublier qu'il y a, dans l'idée de l'art par hasard, l'intention de fuir la nécessité technique sans liberté. Cependant la liberté ne s'identifie pas au hasard. Le hasard, c'est un jeu dans l'instant sans responsabilité. Mais la liberté, c'est un effort dans la temporalité. L'art aléatoire abolit la signification du processus de libération personnelle qui est au cœur de l'expérience esthétique. Car l'expérience esthétique de l'art, soit pour la création, soit pour l'appréciation, implique au fond un engagement personnel, avec l'austérité mortifiante d'un amour ardent et la recherche aventureuse de la liberté, culminant dans l'ascension spirituelle vers les valeurs transcendantes que l'accomplissement de l'œuvre indique symboliquement. Les deux termes opposés de l'expérience, la création et l'interprétation, se situent dans l'horizon de la temporalité. C'est alors le processus qui est important dans l'expérience esthétique.

Si la technique diminue la temporalité et néantise par là la conscience humaine, quels que soient ses fruits pour la vie quotidienne, l'humanité sera exposée au péril de la déshumanisation, si nous la laissons faire, parce que l'humanité ne peut pas survivre sans spiritualité et que la spiritualité est un sommet de la conscience. Nous ne voulons pas dire que la technique soit au fond malfaisante. Non. La vie humaine ne peut s'en passer. Mais la technique est vraiment intoxicante pour l'intériorité humaine, comme nous l'avons dit. Il faut donc résister à la tendance mondiale au technocratisme. L'un des contre-efforts, c'est l'orientation esthétique, car l'expérience esthétique de l'art consiste dans la temporalité, et son essence est le processus vers un but transcendant. De plus l'expérience du beau est vraiment mieux partagée, parce qu'il n'y a pas d'homme qui ne l'ait éprouvée. La tâche de l'esthétique comme calonologie ne se réduit plus à la théorie du goût et à la critique d'art. Elle doit et peut donner l'orientation juste selon laquelle on pourra sauver l'esprit humain du péril technocratique.

LA TECHNIQUE ET L'ART

Sans aucun doute, la technique contribue à la diffusion et à la compréhension des œuvres d'art. On ne doit pas négliger son efficacité considérable pour la reproduction, la restauration, la conservation de l'art, comme pour l'amélioration matérielle de la vie. De plus la technique a élargi le thème de l'art en ouvrant, aux deux extrêmes, un cosmos interplanétaire et un domaine microscopique à notre sens. Ces contributions autour de l'art sont inoubliables. Mais nous disons bien « autour de » l'art. Car les alentours ne sont pas le centre, ils ne sont pas la chose même. On sait bien qu'il y avait une grande activité de l'art lorsqu'il n'y avait pas de technique machiniste. Donc l'art peut survivre sans l'aide périphérique de la technique, bien qu'il y ait des inconvénients à cela. Le fait que la technique apporte une aide périphérique à l'art n'implique pas qu'elle contribue au centre de l'art, à l'âme de l'art. Il faut penser correctement le rapport dialectique entre l'âme de l'art et le progrès de ses alentours.

Considérons un phénomène artistique dans la vie quotidienne d'aujourd'hui : un auditeur écoute à la stéréo un mouvement très beau d'une passion de J.S. Bach en buvant du cognac sur un sofa dans une atmosphère détendue. Ici, il y a reproduction, restauration, maintien, confort, tout ce que la technique nous offre. Mais est-ce de la musique qu'il écoute ? Où est l'âme d'une musique « passion » dans cette expérience ? L'auditeur sait bien qu'il s'agit de ce chœur de la Passion selon Saint Matthieu qu'il aime et il en jouit si bien, dans l'émotion du cognac, qu'il peut à la fin siffler la mélodie sans faute. Et il croit qu'il fait l'expérience de la musique de Bach. Mais ce n'est que l'amusement d'un mélomane qui se divertit à la musique. Et il travestit cette musique. Je répète : où est l'âme d'une musique-passion dans cette attitude ? Cette attitude n'est que jouissance par exagération d'un résultat techniquement abstrait de la musique en entier. C'est une mutilation de la musique : le plaisir sensuel et le beau sont confondus. L'expérience esthétique est nivelée. La beauté devient seulement le plaisir de l'articulation horizontale, la verticalité de l'art disparaît. Les alentours techniques de l'art s'étendent toujours plus, et l'art y perd son âme qui réside au sommet de cette verticalité spirituelle. Si le mot « âme de l'art » ne vous plaît pas, dites aussi bien l'essence de l'art.

Ainsi la technique tout en contribuant au progrès des alentours horizontaux de l'art, peut nuire à l'art. Et nous savons pourquoi : elle diminue la temporalité, qui est essentielle à l'expérience esthétique. Voilà déjà deux raisons pour dire que la technique moderne est l'ennemie de l'esthétique et de l'art. Mais, est-ce bien vrai ? On sait bien aussi qu'il y a de l'art dans ce monde technique d'aujourd'hui et que l'homme y cherche bien l'art. Y a-t-il là une aporie ? Il faut maintenant réfléchir à la relation entre la technique et l'art en considération de leur vecteur et de leur temps.

Le problème du vecteur et celui du temps doivent être éclairés ensemble en considérant la distinction entre l'instant et le *kairos*. L'instant est l'élément du temps. Il y a l'instant à l'infini. Et il y a une série unifiée des instants. Le temps va d'un instant à l'autre dans le cadre de la série unifiée de ses instants. Par contre, il n'y a pas beaucoup de *kairos*. Il n'y a pas de série du *kairos*. Le *kairos* n'est pas un élément du temps, parce que le temps peut couler sans *kairos*. De ce temps sans *kairos*, le vecteur est horizontal et unilatéral ; horizontal, puisqu'il ne coule qu'au niveau de la matière éternellement la même ; unilatéral, puisqu'il coule de l'instant présent à l'instant du terme, de la disparition de la chose. Avant sa ruine la chose fonctionne dans ce temps horizontalement et unilatéralement. Ce temps peut couler sans aucun contact avec l'esprit humain. C'est le monde de l'automatisme, c'est, si la chose est assez bien organisée pour être contrôlée, le monde cybernétique. Telle est l'idée de la technique ordonnée à la chose qui se meut dans un temps fait d'instants. Son vecteur aussi est horizontal, et son temps une agrégation d'instants sans *kairos*.

Mais qu'est-ce que le *kairos*[9] ? C'est aussi un moment, mais ce n'est pas un instant comme dans la série où se meut la chose. Pour mieux l'entendre, citons le mot de Tsau-Chan[10], un penseur chinois du IXe siècle : « Le phénomène est la vérité, la vérité est le phénomène ». Cet aphorisme, loin d'aboutir au phénomène, signifie que dans l'état de l'esprit illuminé ce qu'on voit est déjà ce qui est la vérité éternelle ; le phénomène que l'homme illuminé voit n'est plus en ce monde, mais il est transcendant. Et le *kairos* est justement le moment précis où l'on atteint à cet état d'esprit. Il n'appartient donc pas au temps de la chose, mais au temps du dialogue entre la chose et l'esprit humain. Il est donc le point de contact vertical avec le transcendant. Mais où se fait le temps de ce dialogue ? Autrement dit, comment un homme qui n'est pas encore illuminé peut-il atteindre au *kairos* ? Un moine a posé une question à Tsau-Chan après qu'il a dit : « Le phénomène est la vérité... » : « mais, mon maître, où se manifeste cela ? » Tsau-Chan répond simplement : « Ici », en soulevant un plateau. Réponse énigmatique, comme souvent dans l'entretien Zen-bouddhique, mais non ambiguë. Tsau-Chan veut dire que l'on peut accéder au transcendant par une aussi petite chose qu'un plateau, si on ne la voit pas au point de vue de l'utilitarisme technique, mais au point de vue de la méditation, c'est-à-dire si on ne la voit pas comme une chose mais comme une œuvre. Puisqu'il n'y a aucune chose qui ne soit en relation instrumentale avec d'autres, la chose ne peut pas avoir son monde indépendant, elle n'existe qu'en dépendant des autres dans l'horizon de l'utilité. Au contraire, une œuvre a son monde indépendant, elle est unique en son genre et elle a sa destinée propre. Voir une chose comme une œuvre, c'est donc la voir comme plus que la chose, la voir comme individu, comme un être qui a son fond et sa fin en lui-même. C'est donc chercher hors de l'horizon de l'utilité le principe et la fin de cette chose. Alors la pensée doit se mouvoir verticalement. Et c'est ainsi qu'elle pense le *kairos*, parce qu'il est le point culminant vertical pour toucher au transcendant. Comme disait un

---

[9] Le mot de Pittakos est très important pour entendre l'œuvre humaine. Cf. R. Berlinger, *Das Werk der Freiheit*, pp. 24-36. Et pour la signification décisive du *kairos*, voir mon article : « La mort et l'art comme communication de l'identité » (écrit en japonais), *Tetsugaku Nempo (Annales de Philosophie)*, n° 25, 1963, pp. 1-68, et mon livre : *L'identité et son auto-formation* (en japonais), Tokyo, 1970, p. 420.

[10] Tsau-Chan (ou Tsau-Schan) 839-901. Pour un lecteur européen, il y a le *Zen*, cf. *Aussprüche und Verse der Zen Meister*, Insel Bücherei, n° 798. Il y a beaucoup d'éditions japonaises de Tsau-Chan.

des successeurs de Tsau-Chan, Hsue-Don Dschung-Hsien : « La vérité sainte est la largeur ouverte »[11], là où le transcendant n'est pas caché. Il faut que l'on puisse toujours voir chaque chose comme une œuvre où l'on monte au *kairos* afin de toucher à la vérité transcendante. Mais c'est en effet difficile pour les hommes ordinaires, non-illuminés, parce que la chose en appelle à l'utilisation pour ceux qui s'intéressent à l'utile. Et c'est justement là que l'art trouve sa tâche quant au problème du *kairos*. Une œuvre d'art, on peut la voir comme une chose, parce qu'elle est aussi une chose. Mais nous pouvons sans peine la voir aussi comme une œuvre, parce qu'elle nous ouvre par son charme ou par sa beauté à son individualité. Elle nous appelle au désintéressement, et nous convie à nous placer dans le temps vertical. Ce temps propre au *kairos*, c'est la forme de l'expérience esthétique. La nature de ce temps n'est pas une *dialectique* horizontale entre la chose et un moi universel impersonnel. C'est le *dialogue* entre l'œuvre et ma personne. Il y a 2500 ans, Dsong, un des disciples éminents de Confucius, dit : « L'homme comme il faut rencontre ses amis dans l'art et fait avancer sa charité par ses amis »[12]. Cela veut dire allusivement que l'homme peut entretenir un dialogue spirituel avec des maîtres dans l'expérience esthétique de leurs œuvres d'art et que, par ce dialogue calonologique, il peut se purifier afin d'atteindre au transcendant, qui est la charité en soi. L'œuvre est l'être qui contient la temporalité verticale, dont la culminante est le *kairos*. Son temps n'est pas une agrégation des instants mais la montée dynamique au *kairos*.

La technique coule horizontalement. Son temps consiste dans l'instant. La chose utilitaire est une incarnation de l'horizontalité et de l'instantanéité qui constituent la forme pure de la *res extensa*. L'art monte verticalement. Son temps se règle sur le *kairos*, où se produit l'expérience esthétique. L'œuvre avec sa destinée individuelle et son

---

[11] *Hsue-Don*, 980-1052, *Bi-yen-Lu* (Hekiganroku, Iwanami-bibliothèque 1538, pp. 1-16) ; *cf. Biyen-Lu* traduit par Wilhelm Gundert, pp. 38-49.

[12] Lun Yu, livre XII, 24, V, *The Analects of Confucius*, Londres, 1938. Pour la pensée classique en Chine, on peut recommander : H. Maspero, *La Chine antique*, Paris, 1927 ; M. Granet, *La pensée chinoise*, Paris, 1934. Et mon livre *Betrachtungen über das Eine*, 1968. Après cela j'ai publié un livre *Toy no Bigaku* (en japonais, l'Esthétique en Asie) qui a déjà été traduit en chinois. Ce livre contient beaucoup de matériaux nécessaires à l'étude de la philosophie chinoise.

interprétation[13] personnelle est une incarnation de la verticalité et de la temporalité qui constituent la forme pure de la *res cogitans*[14].

Alors la technique et l'art n'entrent pas en collision l'une avec l'autre, parce que leurs vecteurs ne se situent pas au même niveau. Mais il y a naturellement une tension forte entre l'horizontal et le vertical. Cette tension est destinée à l'homme, comme l'existence démiurgique au domaine de la matière qui est commise à la liberté humaine. Bien qu'il n'y ait pas de collision entre les deux, puisqu'il y a cette tension, l'aide mutuelle est par principe impossible. On ne peut penser à la relation entre la technique et l'art de façon optimiste. Jusqu'à présent, on a pensé souvent que les deux sont identiques en tant que « faire ». Mais l'*homo opifex* n'est pas le Dieu pour qui seul l'art est la technique, et la technique est l'art, parce qu'il crée chaque chose comme son œuvre dans une éternité où il n'y a pas de temps. Pour nous, hommes, nous vivons dans le temps, partagés entre l'horizontalité et la verticalité. L'engagement entre la technique et l'art les compromettent toutes les deux. L'aide mutuelle, s'il y en a une, joue toujours dans le domaine des alentours et aussi par hasard. On ne peut pas faire un poème quand on manœuvre son auto, c'est sûr ; on peut pourtant voir des arbres voler derrière la vitre à toute vitesse, et ce spectacle peut devenir une matière pour la poétisation, mais ce n'est pas sûr : c'est un accident.

On peut schématiser cette tension du temps entre la technique et l'art.

---

[13] L'interprétation n'est pas passive. Au contraire, c'est aussi une montée active de l'esprit. L. Pareyson l'a très bien montré dans sa conférence au IVe Congrès international d'Esthétique 1960 à Athènes. Voir les Actes de ce congrès, Athènes, 1962. L. Pareyson : *De l'interpretatione*, Milano, 1964 ; T. Imamichi : *Position logique et orientation de l'interprétation* (en japonais), Tokyo, 1970.

[14] P. Hindemith écrit de J.-S. Bach : *Ist es einer Musik gelungen, uns in unserem ganzen Wesen nach den Edlen auszurichten, so hat sie das Beste getan. Hat ein Komponist seine Musik so weit bezwungen, daß sie dieses Beste tun konnte, so hat er das Höchste erreicht. Bach hat dieses Höchste erreicht.* C'était la conclusion du discours de P. Hindemith à la fête en l'honneur de Bach à Hambourg en septembre 1950. Voir, Paul Hindemith, *Johann Sebastian Bach*, Insel-Bücherei n° 575.

	La technique	*L'art*
L'orientation.	La coulée horizontale par instants.	La montée verticale. au *kairos*[15] .
Le sujet.	La pluralité impersonnelle.	L'individu personnel.
La fonction.	Le contrôle automatique du monde de la chose.	La purification de l'esprit par l'expérience esthétique.
La méthode.	La transfiguration de la matière pour l'utilité.	La cristallisation d'une œuvre et son interprétation.
Le but.	L'autisme de la machine.	L'extase dans le transcendant.
La valeur.	L'efficacité.	La beauté.

L'actualisation d'une chose se réalise dans sa fonction avec sa méthode propre. L'actualisation ou l'acte de l'art se réalise ainsi dans sa fonction avec sa méthode propre, c'est-à-dire, selon ce schéma, dans la purification de l'esprit par l'expérience[16] esthétique avec la cristallisation d'une œuvre et son interprétation.

Ainsi l'expérience esthétique invite à l'extase[17], à une temporalité dont le terme est l'éternité, alors que la signification de la technique est l'autisme de la machine.

ÉTUDE COMPARÉE DE L'EXPÉRIENCE ESTHÉTIQUE EN OCCIDENT ET EN ORIENT

Dans la tradition esthétique de l'Occident, à savoir, dans l'origine grecque classique, il y a eu deux théories éminentes de l'expérience esthétique ; la première est la théorie de l'extase comme enthousiasme platonicien pour la création artistique, c'est-à-dire pour l'expérience

---

[15] *Cf.* mon article cité, notes 6 et 9.

[16] Le mot expérience ne veut pas seulement dire le vécu comme Erlebnis, mais aussi l'objectivation de l'intellect, à savoir Erfahrung, qui suggère aussi : sortir au delà du monde matériel, parce que le mot allemand « fahren » veut dire « se déplacer par roues » afin de sortir de chez soi vers l'endroit voulu.

[17] L'« extase » diffère de la fascination. L'extase vient du mot grec « ἔξστασις », c'est-à-dire ex-stasis, à savoir « la transcendance de l'esprit qui se libère d'une situation fermée », tandis que la fascination n'exprime que le ravissement de l'esprit face à un phénomène ou une chose matérielle.

esthétique positive créative, tandis que la deuxième est la théorie de la catharsis comme purification aristotélicienne pour l'expérience esthétique passive de jouissance. Sans création artistique il n'y a pas de chef-d'œuvre artistique, dans lequel l'esprit humain peut parfaire l'expérience de la catharsis. Sans catharsis de l'esprit humain il n'y a pas de manifestation artistique de l'art comme chef-d'œuvre. Création et catharsis sont les deux phases splendides de l'expérience esthétique selon la tradition occidentale.

De manière tout à fait parallèle, dans la tradition esthétique de l'Orient, à savoir dans la tradition chinoise classique, il y a eu deux théories éminentes de l'expérience esthétique ; la première est la théorie de l'extase comme transcendance confucéenne pour la purification de l'esprit humain, c'est-à-dire pour l'expérience esthétique interprétative de jouissance, tandis que la deuxième est la théorie de la libération extatique comme élévation créatrice de Tschuang-tschou. L'interprétation verticale de la perception de la surface de l'œuvre artistique de l'idée du beau et du bien chez Confucius, ainsi que la création par l'ascèse pour la libération chez Tschuang-Tschou, sont les deux phases splendides de l'expérience esthétique selon la tradition orientale.

Que ce soit en Occident ou en Orient, l'expérience esthétique a deux phases, soit l'expérience productive créatrice par laquelle l'œuvre d'art surgit de la matière et l'expérience de jouissance de l'œuvre d'art par laquelle la splendeur du beau surgit de la forme de l'œuvre d'art. Ce sont les relations subjectives de l'homme envers l'art comme objet ou corrélat de son expérience. Quant à l'essence de l'expérience esthétique concernant l'art, on retrouve alors un parallélisme dans les deux parties du monde.

Cependant il y a une différence essentielle entre les deux traditions, quant à la méthode permettant d'amener l'expérience esthétique à son sommet. En Occident, il y a deux sortes de succès pour la création ; la première est l'enthousiasme platonicien[18] et la deuxième est la méthode poïétique aristotélicienne. Selon la première on ne fait rien, on doit seulement attendre ; selon la deuxième, on doit étudier la

---

[18] Dans le dialogue *Phèdre*, Socrate dit que « τὰ μέγιστα τῶν ἀνθρώπων γίγνεται διὰ τῆς μανίας ». Et dans le dialogue *Ion*, Socrate utilise le mot « ἐνθουσιασμός », c'est-à-dire « être dans le Dieu ou la divinité » pour la cause de l'activité poïétique d'Homère.

méthode systématique[19]. En tout cas, il n'y a pas d'ascèse religieuse. Et dans l'expérience de jouissance, c'est-à-dire la catharsis, on ne fait rien positivement, on doit seulement jouir de l'art selon son dynamisme afin d'acquérir la *catharsis*. Alors ici non plus, il n'y a pas d'ascèse religieuse faite subjectivement.

Mais en Orient, selon l'expérience esthétique de jouissance comme interprétation confucéenne[20], on doit essayer de transcender la limite logique de la pensée scientifique et pour cela, comme nous l'avons montré dans une étude spéciale consacrée à l'esthétique chinoise, on doit absolument s'engager dans la liturgie religieuse pour le Ciel. Et dans l'expérience esthétique créative comme libération *tschuang-tschouenne* du monde phénoménal pour la production du chef-d'œuvre artistique, comme nous l'avons exposé dans l'étude en question, on doit se soumettre à une ascèse tout à fait austère. Que ce soit pour l'interprétation ou pour la création, ici en Orient, on ne doit pas séparer l'expérience esthétique de l'art et l'expérience ascétique de la religion.

D'où le concept métaphysico-religieux de *dao* (selon la prononciation japonaise *dô*), *michi* en japonais, qui veut dire la voie ; celle-ci est non seulement métaphysique, mais aussi absolument religieuse. C'est la clef de l'expérience esthétique selon la tradition orientale.

SIGNIFICATION DE L'EXPÉRIENCE ESTHÉTIQUE ET DE L'ESTHÉTIQUE COMME CALONOLOGIE

Dans la tradition esthétique de l'Orient, et spécialement au Japon ou en Chine, il y a un mot central, *michi* ou *dô* en japonais et *dao* en chinois[21],

---

[19] La méthode systématique qu'utilise Aristote est la méthode de la catharsis comme purification du pathos.

[20] Nous avons exposé, dans ses grands lignes, la théorie esthétique de Confucius dans notre conférence aux Entretiens de l'Institut international de philosophie à Kyoto en 1994. On trouvers dans les *Actes* la version anglaise de notre texte. Voir également notre livre *Studia comparata de aesthetica* (écrit en anglais, français et allemand), et en particulier notre texte « Aesthetik in der chinesischen klassik » dans lequel nous avons exposé les théories de Confucius et de Tschuang-Tschou.

[21] La langue chinoise et la langue japonaise sont des familles linguistiques tout à fait différentes. Mais quant à l'écriture il y a une affinité étroite entre les deux. Les Japonais n'avaient pas de lettre jusqu'au moment où ils utilisèrent les lettres chinoises. Par exemple pour le mot japonais *michi* (la voie), on utilise le caractère chinois *dao* qui

qui veut dire la voie. On en fait au Japon le mot *Gei-dô*[22] qui veut dire la voie de l'art. Ce n'est pas identique à *Yû-gei* qui veut dire le jeu de l'art, le divertissement ou le goût. *Gei-dô*, la voie de l'art, c'est la voie sur laquelle l'*homo viator* marche. L'expérience esthétique est alors le processus d'une catharsis spirituelle, la voie préparatoire pour le salut même dans ce monde technologique, où l'homme reste inquiet et veut toujours être plus libre.

Pourquoi l'homme est-il toujours mécontent, malgré l'efficacité de la technique ? Pourquoi l'homme demande-t-il toujours un « plus » à la technique, même quand il en a assez ? Est-ce à cause d'une volonté de progrès pour l'humanité ? Non, parce que l'humanité n'existe pas seulement pour l'efficacité. Ne faut-il pas s'arrêter un moment pour chercher si le progrès ne concerne pas aussi l'intériorité ? Mais on veut ne pas s'arrêter un instant. On veut avancer toujours plus vite sans aucune considération pour la dignité de l'intérieur. Pourquoi ? Parce que l'on veut oublier son angoisse. Pourquoi cette angoisse ? Parce que le problème le plus grave, la mort, atteint l'homme à travers la cuirasse de la technique et le blesse plus profondément qu'autrefois, parce qu'il a perdu sa vigueur spirituelle.

Cependant, si l'expérience esthétique[23] autorise toujours l'extase, cela veut dire que l'esprit humain peut vivre seul en se détachant du corps. La mort n'est-elle pas la grande extase de l'esprit au-delà du corps ? L'expérience esthétique, c'est un petit mystère de la mort. L'esthétique est alors l'ontodicée, qui doit constater la possibilité de l'horizon de l'être éternel, que la technique veut remplacer par sa fonction continue. L'œuvre d'art attire l'âme. L'expérience esthétique,

---

signifie la voie, et alors on doit prononcer ce caractère chinois comme *michi*, si on veut parler en japonais. Mais en même temps, les Japonais ont appris beaucoup de concepts chinois qu'ils n'avaient pas eus dans leur histoire, par exemple *Yi*, qui veut dire « responsabilité ». Dans ce cas, les Japonais ont dû imiter la prononciation chinoise afin d'apprendre ce concept. Nous Japonais, prononçons pour cela *gei* et non *yi*. De toute façon *michi* et *dao* signifient tous les deux la même chose : voie.

[22] Le mot *Gei* est aussi chinois et il indique vraiment l'art. Jusqu'au IVe siècle av. J.-C. les Japonais utilisaient, pour le mot art, *takumi* qui veut dire « skilfulness » en anglais. Après le VIIe siècle, le concept *Gei* (art) est devenu familier au Japon.

[23] M. Dufrenne a indiqué l'importance de la signification ontologique de l'expérience esthétique, v. M. Dufrenne, *Phénoménologie de l'expérience esthétique*, Paris, P.U.F., 1953 (2e éd. 1967), 4e partie, « Signification ontologique de l'expérience esthétique », pp. 657-677. La première version de cet essai fut écrite originellement pour les *Mélanges* de mon ami Mikel Dufrenne.

quant à son degré primitif, peut être engendrée dans l'âme passivement. Elle est ouverte à toutes les âmes. Elle peut surgir partout même dans le temps technique afin de changer cette âme attirée dans l'esprit qui transcende. Dans ce temps, sans expérience esthétique, on n'a pas de voie pour la transcendance de l'esprit, et sans esthétique on n'a pas d'occasion pour penser à l'éternité.

POLITICS

Francisco MIRÓ QUESADA C.
University of Lima

1. TERMINOLOGICAL CLARITY

The aim of this essay is to speak, from a philosophical point of view, on the present political problems, national and international, that are endangering the democratic organisation of modern societies. To achieve this goal, these problems must be rationally faced and analysed. I intend to cope with them in order to see if some favourable issues can be found. But before beginning our investigation, we must have a clear idea on the meaning of the word « politics ». Unfortunately, this meaning is vague, and the reason for this lack of clarity is due to the fact that politics consist in human actions practised through time and space and, consequently, the concept of politics is empirical[1].

When we have to handle empirical meanings, we must take into account that they always have two components : 1) a nucleus of clearness ; 2) a margin of fuzziness. This means that, concerning empirical concepts, a twofold situation is always present. In many cases it is very easy to apply a term to the objects that belong to its extension. For instance, if you want to refer to objects that belong to the extension of the word « chair », you can easily find things that are unmistakably chairs. However, there are also many objects that cannot be classified with certainty as chairs or non-chairs. For instance, all chairs must have a back support. But when this support is too small, it is impossible to know if it is a chair or a bench. When the term is the name of a simple material object, things are not so difficult ; but when they express complicated empirical structures or dynamisms of social phenomena, they may be quite unmanageable.

[1] In what follows, to avoid repetitions and circular descriptions, I shall use the expressions « term » and « word » as synonymous. The word « concept » shall be used as a primitive term. I presuppose the reader has a sound intuitive understanding of its meaning.

P. Kemp (ed.), World and Worldhood, 169-186.

2. AN INTENT OF DESCRIPTION

Since the times of Plato and Aristotle many philosophers have tried to find a universally acceptable definition of politics. Let us try to capture the common distinctive traits that can be found in some present well known authors. A universal trait that is present in every definition or description of politics, is that political activities are concerned with power. According to Weber, politicians are motivated by the will to conquest power, to share it, to distribute it, and to bargain with it (Weber, 72). Duverger says that politics, as an institutional organisation, is the endeavour to conquest power, and also to reach integration (Duverger, 67). Easton considers political activity as a kind of feed back between governmental power and the reaction of the individuals, groups and institutions that want to orient this power with respect to their interests (Easton, 69). According to Parsons, politics are concerned with power and power is a generalised medium in which the political process follows its course (Parsons, 69). For Perroux politics is a set of relationships among individuals and groups motivated by contradictory trends, and trying to establish stable conditions of authority and dependence (Perroux, 64, 69). Francisco Miró Quesada R. says that politics is a complex network of power relationships which has institutional, conductive and ideological expressions (Miró Quesada R, 94). There are other authors that deserve to be quoted, for instance, Lucas Verdu, Dahl, Deutsch, Toffler, Van Dyke, Yakovlev, *et alia* ; but I think that the ones we have cited are more than sufficient to show that the concept of politics is constitutively related to the concept of power.

Trying to apprehend the common features of the former definitions, I risk the following description (not a definition) : politics is a human activity exerted in order to conquer power or to maintain it, and having collective effects. Of course, this struggle is influenced by a myriad of social structures and dynamisms. For instance the way a conservative person looks at political elections, greatly differs from the vision of a leftist. But both of them, although in a quite civilised manner, are fighting for power.

3. POLITICS IN PREVIOUS TIMES

International policy in ancient Greece and Rome was a permanent fighting for power. Of course, it also consisted in gaining acceptability, in making pacts and alliances. But the principal means to win a confrontation between two city-states, was *war*. War was seen as a most natural thing. And, although common people complained about its tragical consequences, no one thought that universal peace was a goal attainable without military subjection of the rival countries. War was made by city-states that were small nations, or among internal national groups. When a nation (a city state or a group belonging to some nation) lost the war, it was perfectly natural to kill most of the warriors and to sell as slaves the remaining ones, together with the women and children. When the fight was internal between two rival leaders, in most cases, defeat meant death and, not a few times, torture.

In the Middle Ages the situation was not so desperate, but the struggle between two nations or between two feudal lords could be also very cruel. During the Renaissance, although intellectual and courtesan life was more refined, charity was the exception.

But in the XVIIth century cruelty began to decrease. I think there were two reasons for this change : the political and military power of the Catholic Church began to decline, and the ideal of rational life, through a rather complicated process, began to be diffused through constantly growing groups[2]. This diffusion had noteworthy consequences. The most important was the flourishing of a new type of intellectuals which dubbed themselves « libres penseurs ». Freedom of thought became so highly appreciated that some absolute monarchs began, although within certain limits, to tolerate it, and even to have some sort of snobbery about showing off their learning and their personal friendship with the most famous intellectuals.

Wars were still cruel but something really new began to happen in Western Civilization. For the first time in history, important thinkers like Leibniz and Kant began to consider that wars, after all, were not unavoidable. And they even wrote about the possibility of establishing universal peace through international treaties.

4. POLITICS IN THE XIXTH CENTURY

But these noble proposals of intellectual giants, although well received among bourgeois groups, could not fully overcome the traditional attitudes of the political world. The enlightened monarchs, when they had to face problems of power, forgot all about the doctrines of the philosophers they pretended to admire. All of them, surrounded by ambitious militaries, adulating courtisans and adventurers, motivated by

[2] I do not say that the Christian Church, Catholic or non-Catholic, was in itself cruel, but only that material and political power lead astray any kind of church from its spiritual mission. The struggle for power is quite contrary to love and compassion that are the common ends of all superior religions.

About the process that leads the Western civilization to the adoption of the rational life ideals, see Miró Quesada C., 87.

greed of richness and eagerness of glory, became involved in bloody wars. And the great progress of technology, due to the Industrial Revolution, allowed a rapid and systematic expansion of imperialism. On the other hand, a boundless growing of nationalism during the XIXth century, had desastrous warlike consequences in Europe. A great majority of citizens of the most advanced countries, sincerely believed that war was justified if it was for the greatness of their nation. War, in order to conquer a foreign land, was applauded by the whole population. When the army that was to subdue a nation, paraded through the streets, with battering drums and flaming banners, beautiful girls threw over them delicate flower petals.

This attitude towards warfare was quite manifest until the middle of the XXth century. Unbelievable, as it sounds nowadays, a few years before World War II, to have colonies was considered the most natural of things. When Mussolini attacked Ethiopia, France and the United Kingdom bitterly criticised him, but they never intended to set free their own colonies. However, things had begun to change among certain advanced political groups. For the first time in Western Civilisation, pacifism was transformed into an organised movement.

5. A 180 DEGREES REVOLUTION

Suddenly, things completely changed. War began to be seen, instead of a glorious achievement, as a shameful activity. When two nations were involved in war no one accepted to be the aggressor. Aggression was a sin, only defensive war could be accepted. Once the Organisation of the United Nations was born, when two nations began a war, parallel to the bloody conflict, there was an unending trial in which the quantity of the proves presented was only comparable to the astonishing complexity and messiness of their meaning. Peace was now, not only desirable, it was the highest duty. Every non-defensive war was an abomination. History was considered as the secular march of humanity toward universal peace. In a few years international moral values were radically transmuted. Although there still was a great difference between value recognition and political international action, even the most cynical violation of peace was always accompanied by an official moral justification.

Why did this change happen? As is always the case in human affairs, causes and motivations are dreadfully complicated. Of course an immediate reason was the monstrous slaughter of World War II. But it was not the only one. A deeper cause was the change experienced by the conception of man. This change was due to the *ideal of rational life*. A direct consequence of this ideal is that every human being must be considered as an *end in itself* and never as a *means or instrument*. This consequence was, for the first time, enunciated with explicitude by Kant, in the last third of the XVIIIth century[3]. I call this Kantian principle : « *rational humanism* ». In this kind of humanism, the fundamental relation between rationality and morality is enunciated with full clarity[4].

Rational humanism was, at first, adopted in philosophical and, generally, in the « intelligentzia » circles. Through a slow process of irradiation the rational life ideal was making its way. The « Aufklärung » of the XVIIIth century oriented the regular citizens towards new values and new ethical criteria. Although during the XIXth century, conquering wars were still approved by common people, in learned milieus the pacifism was preached with increasing influence. During the First World War many intellectuals, and some politicians, strongly opposed the chauvinism of most politicians and the majority of the population. The famous philosopher Bertrand Russell was imprisoned because he publicly declared that the war was a stupid feud among kindred royal families. And the influential socialist politician, Jean Jaurès was assassinated for being a pacifist.

The genocide of the Second World War was so atrocious, that the humanist ideas that had been slowly but firmly advancing, suddenly embraced the whole world. As a result of this process, the national chauvinism that had justified wars and colonialism, could no longer be accepted. The Kantian rational humanism was now explicitly or implicitly recognised by all the nations of the Earth. This acknowledgement was the source of the creation of a fundamental international institution : the United Nations Organisation. A result of this transcendental historical fact, was the Chart known as the *Universal Declaration of Human Rights*. The Article 1 is paradigmatic :

> *All human beings are born free and equal in dignity and rights They are endowed with reason and conscience and should act towards another in the spirit of brotherhood.*

---

[3] Kant was anticipated by several philosophers, ancient and modern, for instance, Seneca and Marcus Aurelius in Rome, and Spinoza in the Seventeenth century. In the last part of his *Ethica-More Geometrico Demonstrata* and, especially, in his *Tractatus Teologico-Politicus*, Spinoza deploys a humanist conception of society. But, from the principles and definitions of the *Ethica*, does not clearly follow the moral engagement to contribute to the humanisation of history. In the *Tractatus* this engagement is fully exposed. Nevertheless I think that Kant develops his conception of rational humanism and its meaning for history, in an unsurpassable way.

[4] Kant did not use the term « *humanism* », but his moral doctrine has all the traits of rational humanism. The word « *humanism* » was used, for the first time, in 1808 by the German philosopher of education F.J. Niethammer. Kant had died in 1804 and I think that Niethammer's term was inspired by the *Kritik der praktischen Vernunft*.

This article is full of meaning. The words « reason », « conscience » and « brotherhood », are all embracing. « Reason » has, no doubt, a philosophical meaning : it is a clear mention of the ideal of rational life. « Conscience » refers to the fact that we all have an intuition of what is good and what is evil. It presupposes the existence of natural rights. These rights can be determined by rational intuition, an intuition that was foundational according to the criteria of the great rationalist philosophers of the Seventeenth century, like Spinoza and Leibniz. And « brotherhood » has a religious meaning. If we are the sons of God, then we are brothers.

From the first article can be logically derived many important consequences. And the same can be said concerning the other articles. But, besides this possibility, there is another very important one : if we take the first article as a premise then, the other twenty-eight articles of the Chart can be derived as consequences of the first. But to do this, besides classical logic, we must apply a kind of circumscriptive or non-monotonic logic. This thesis may seem strange to the layman and even to most philosophers of law. But I think that it is very important to point out the fact that there exists a deep and total imbrication concerning the twenty-nine articles of the Chart. Of course, I am not saying that the twenty-eight remaining articles are supernumerary. They are pedagogically necessary. But I want strongly to dwell on the essential factum that the Universal Declaration of Human Rights would never have been possible without the deep faith of us, modern citizens of the World, in the effectiveness of reason to face the theoretical and practical problems that concern our terrestrial existence.

6. FROM PRINCIPLES TO ACTION

My point of departure to handle the philosophic-political problems which are characteristic of the final years of our century is, now, settled : it is *rational humanism*. The deep foundation of this kind of humanism is, as we have seen, to consider every human being as an end in itself. This fundamental fact can be expressed by what I call : the « *autotelic principle* ». An immediate consequence of this principle, is that that we must endeavour to transform history, to surpass an irrational process to transform it into a rational one. And this endeavour is tantamount to the permanent effort to base history on dignity, freedom and democracy.

Sceptic philosophers affiliated to pragmatism, relativism or historicism, will adduce that the searching of means to fulfil concrete aims, has a clear meaning. But to speak of « ends in themselves » is too obscure to be taken as *the* fundamental principle. A second argument is that, even if we knew what reason is all about, this knowledge would be useless, because most people are not led by rationality but by egoistic irrational motivations. Consequently, the rational life ideal is only a pretext to dissemble our true intentions and to justify our maneuvers to increase our individual or national power. Both arguments are confutable as we shall now prove.

7. REASON

7.1 Against vagueness

It is true that the word « reason » has a myriad of meanings. But although some of them are irretrievably vague, many are quite clear when used in the appropriate context. This context is superbly expressed in the first article of the Universal Declaration of Human Rights. In this declaration the word « *reason* » is employed as a necessary condition of political international and intra-national behaviour. A political action that is not based on reason, cannot be accepted, neither by the international community, nor by internal national policies. So, « *a contrario sensu* », a behaviour that is contrary to reason cannot be justified. In the context of the Chart, the word « reason » has a very clear meaning : *non-arbitrariness.* To this interpretation, the sceptik immediately objects, by saying that « arbitrary » is still vaguer than « reason ». The concept of *arbitrariness* is empirical, and because of this, it has a considerable margin of fuzziness. Yes, but it also has a precise nucleus of transparency. To show this, it is sufficient to recur to ordinary language. Let us look into some well-reputed dictionaries. In the Oxford Dictionary we find that « arbitrary » means : derived from mere opinion ; capricious ; *unrestrained, despotic.* In the Diccionario de la Real Academy Española, « arbitrary action » means an action against justice, law or *reason.* And in the Larousse Universel, « arbitrary » is a behaviour which depends on the sole will, without any consideration of *reason,* equity and law.

Now, there is nothing easier than to describe the behaviour of a tyrant. There are, of course, some actions that cannot be qualified with precision. But an overwhelming majority of cases can be recognised, with full certainty, as being despotic. It is not a mere coincidence that the word « *reason* » is included in the quoted dictionaries. This inclusion is very important, because it shows that *reason* is tightly related with *justice* and *law*. It is obvious that *reason* is the antonym of *arbitrariness. Non-arbitrariness is a constitutive trait of rationality.* Once the mentioned traits are determined, it is possible to find their logical relationship. The trait of *non-arbitrariness,* is, as we have seen, constitutive regarding the concept of *rationality.* This means that the intension (with « s », also called « comprehension ») of the concept of rationality, *includes* the concept of *non-arbitrariness.* And this inclusion means that the trait of non-arbitrariness can be *logically* deduced from the concept of rationality. That is, the relation between the two concepts is analytical. But the importance of this relationship exceeds its moral meaning. As a matter of fact, *non-arbitrariness* is also a necessary condition of scientific and philosophic knowledge. It is a « *contradictio in terminis* » to say that *rational knowledge* can be constituted by mere whim or by the imposition of a legal system.

But there is other trait that is also a constitutive note of rational human behavior : *symmetry.* An impressive example of this meaning is to be found in the first and second articles of the Universal Declaration : « All human beings are born free and equal in dignity and rights » ; « Everyone is entitled to all the rights and freedoms set forth in this Declaration ». Let us give a simple illustration. Suppose C has lent money to D and, according to the contract, the debt must be cancelled in time T. If D does not return the money in due time, then C is entitled to suit D for default. Let's suppose, now, that after some time it is C who is in need of money, and that D lends money to him, with the same stipulations regarding maturity. Then, if C refuses to pay D in due time, D has the right to resort to the tribunals to obligate C to give the money back. Generalising : the duties and rights that can be established between C and D, are symmetrical. In shorter terms, if C, in situation S can enforce D to assume conduct E then, in the same situation S, D can enforce C to assume conduct E[5]. It is absolutely evident that a non-symmetric system of legal norms cannot be just. The same considerations are valid regarding moral norms[6].

7.2 Pro reason

It is not true that human beings cannot behave motivated by noble sentiments and rational ends. In the human world we find every kind of behaviour, from the vilest to the noblest. But what I intend to show is that, in present times, most of the laws that regulate human behaviour

[5]The analogy, of course, cannot be perfect because some time has elapsed between the loans of C and D. And perhaps some illegality is discovered in the behavior of D, or a strong process of monetary inflation has taken place between both loans, etc. But the fact is that, in normal situations, the duties and rights of C and D, must be the same. And even in abnormal ones. Although it can be quite difficult in some situations to attain symmetry, the principle of symmetry is always a goal to be achieved within human possibilities.

[6] When we speak about non-arbitrariness and symmetry we must be aware that they are complementary principles. Non-arbitrariness is a *necessary* condition of justice, and symmetry is a *sufficient* one.

comply the conditions of non-arbitrariness and symmetry. An outstanding example that we have already exposed, is the Universal Declaration of Human Rights. And in most constitutions of the nations that adhered to the Universal Declaration, or who joined the United Nations Organization after the proclamation of the Chart, both principles are explicitly or implicitly included. An example, among many, is the Peruvian Constitution. In its first article, it says :

> *The protection of the human person and the respect of its dignity are the supreme ends of society and of the State.*

If the protection of the human person is the supreme end of society and state, then human beings cannot be treated as instruments but as ends in themselves. This means that the laws of a nation must be non-arbitrary and symmetrical. Because if they are not, the citizens can be treated as mere instruments by the person or party group that hold the political power.

Nevertheless, a sceptic can defend himself by adducing that even when a legal system can contain non-arbitrary and symmetric norms, considered in its entirety, it can be arbitrary because it can include some institutions that violate the autotelic principle. The reply to this, is *history.* We shall explain this answer in what follows.

8. A TASK FOR PHILOSOPHY

Once scepticism has been overcome, we become conscious of the most important task that philosophy has to perform in present days. It is unquestionable that the grave problems that threat mankind today, cannot be solved without a determined *Weltanschauung* and, very specially, without an organized system of values. For instance, why must we be pacifists ? Some people think that we could live better than we really do, if some human groups could be eliminated. Why must we reject racism ? Why must we respect the Universal Declaration of Human Rights ? From a practical point of view it might be convenient to oppose racism ; but from the same point of view it could also be convenient to commit a genocide. The only way to avoid these purely pragmatic way of seeing the world, is through philosophical reasoning. But this reasoning must be objective, so that all men and women that are capable of thinking can be convinced. Otherwise philosophising would be worth nothing.

According to our previous philosophical set up, the political problems must be selected and settled into a hierarchy according to the following criterion : the grade of hindrance they present to the fulfilment of the rational life ideal. For instance, there is no doubt that the international power and wealth relations among different countries are asymmetrical ; and that, in many countries those that rule power are arbitrary. As far as the « developing countries » are concerned, arbitrariness and assymmetry, so much in the international relations as within the intra-national political organizations, are strongly rejected by the majority of the population[7]. There are, of course, as it is always the case, exceptions, but these are limited to the same nations (not to all of them) in which the cultural level of the population is very low.

Although there are many ways to classify objects, I think the simplest one is dichotomic division. This kind of divisions can be made in many ways. I choose, naturally, the dichotomy that, as I have anticipated, is the most convenient concerning symmetry and non-arbitrariness. From the point of view of nationalities, problems are external or international, and internal or national. From an international point of view, the gravest problems are the assymetry of power among nations, and the arbitrary behaviour of the mighty ones. From a national point of view the most dangerous are the assymetry of power concerning social classes, and the arbitrary behavior of the highest ones. It is important to be aware that arbitrariness and assymmetry are specially felt in educational systems.

9. NATIONS AS BASIC PROBLEMS

9.1 The arbitrariness and assymmetry of power

From the most remote and obscure times, up to the present ones, there have been dominant and dominated nations. If we apply the criterion of non-arbitrariness and symmetry, we find that international life has been characterized by a peculiar tension between advances to, and draw backs from, these two goals. But, in the long run the autotelia has emerged reinforced. Some times with a feeble advantage, some other times through accelerated, even revolutionary processes. The advance goes forth and back, and then forth again. It cannot be predicted with certainty, but it is not illusory. History seems to advance from the utmost

---

[7] Symmetry in wealth distribution is a very deep and difficult concept to analyse, that must be studied with care. In a capitalistic economy it is impossible to reach perfect symmetry. But it has been demonstrated by several authors (for example, in the fix point theorem of Kakutani) that perfect competition produces a situation in which capitalistic profit is equal to zero. This means that the market prizes are the lowest that can be attained. And a society in which this happens cannot be called capitalistic. The result of this imaginary economy would be the limitation of assymmetry to salarial differences. And this is completely anomalous regarding the natural expectations of investors who, naturally, expect revenues from their investments.

assymmetry of the slavery ancient society, toward a growingly approach to symmetry (full democracy).

9.2 The danger of universalism

But there is a danger in this process. If its goal is the forging of a rational humanist society, then its principles must be universal. As a matter of fact, if non-arbitrariness and symmetry had not universal ends, there would be a kind of self contradiction in the conception of rational humanism. If we gaze at history as a millenary process, there is an obvious march toward universality. This march is destroying most of the historical value tables, including our own. The local customs are changing with dazzling acceleration. The advance is threatening our sense of homeliness. But we, human beings, need to be settled in a concrete territory. Patriotism is the deep expression of this parrochial need. For this reason, many people see with fear or with rage this dissolving process, in which selfidentity may be lost. The result of this process is *separatism.* This situation is really dangerous, because the march to universality is the march of reason, whereas the way of parrochialism leads to irrationality and, consequently, to arbitrariness. Neo-nazis are appearing in some European countries, racism is beginning to grow with a dangerous rate. Groups of people that, until recently, were members of a nation, suddenly feel the necessity of separating from the majority and to forge an independent society. For instance, the Bretons and the Corsicans in France, the Basques in Spain, the francophone citizens of Canada. The true significance of separatism is revealed by the fact that the majority of these minorities do not want to be separated from their respective countries. So, the minority feels itself non-understood and even betrayed. And they resort to the only issue that is left to them : terrorism.

The only way to supersede the parrochialist danger is through authentic universalism. This issue may seem paradoxical, but it is not. On the contrary, it is the only way to tolerate parrochialism and to avoid its corrosive force. Because a true universalist has surmounted the need to have a private territory ; he is a citizen of the universe, he belongs to the cosmic race. So, when a part of his own country manifests separatist tendencies, he is quite disposed to accept the partition. Universality can only be achieved when it is able to include the concrete differences which fall under its scope.

9.3 Education

The present and multiple tendencies regarding separatism give the impression that the only way to overcome its danger is through intense education based on universalist principles. Yes, but to believe that the faith in universality is sufficient to solve the problem of separatism, is simply wrong. Because separatism can only trouble people that with utmost good faith consider themselves to be universalists concerning human values. However, they are not true universalists, they are still prisoners of the blind and aggressive nationalism that was the scourge of Europe and other regions of the world during the XIXth century and the first half of the present one. This nationalism was so strong and obnoxious that its effects, as we have just said, are clearly recognisable in the reaction of many so-called « universalists ». For them, separation of a territory from the national unity, is still a sin.

But the true full conscious universalist is not afraid at all by separatism. Quite the contrary, he considers it with approval. Because, by sheer logic, universality must contain particularities. A civilization is universal when it has determined qualities that are shared by all the human groups and all the nations that belong to it. If a group of some country wants to be independent, the only rational way to react, is to help them to attain their goal in a minimum of time and with a maximum of efficacy. Because the universalist knows that, once the group is separated, it will keep the universal values of humanism and democracy. And there is also the possibility that when the group finds that he can reach its goal with such extreme facility, then its eagerness for separation could, perhaps, disappear. This second alternative happened in Spain with the Basques and the Catalans. Most of them saw with full sympathy the separatist movement. But when separation was submitted to a referendum, the immense majority voted for union[8].

9.3.1 Assymetry of power

The assymmetry of power within the population of a nation is the consequence of many factors. The most notorious are the division of labour and the lust for power. Division of labor is necessary to face the problem of scarcity. When the population begins to grow, the alimentary

[8] Of course difficulties can can always be found. For instance, there can be the danger that, once it is separated, the group will impose cruel penalties on its subdits, or that a political movement will succeed and, once installed in power, will impose a totalitarian regime, etc. Yes, all this is possible but it has never happened. The case of Islamic fundamentalism has nothing in common with separatism because it has always been imposed by great majorities.

needs augment very rapidly. The only way to face this problem is by increasing the productivity, and this augmentation cannot be obtained without organising the methods of sowing and harvesting. Moreover, men must devote themselves to hunting and defending their family from the attacks of wild animals and rival tribes. And the difference in physical strength among men and women lead, in a natural way, to maintain women limited to household work and agricultural care. This situation is the pre-historical source of patriarchal society.

Most anthropologists agree on the fact that in the dawn of tribal life, there were rival groups that fought for power. Through this fight and the division of labour, social classes were formed. And, through a complicated social and psychological process, power was accumulated in the hands of a minority. As we have said, the distribution of power became totally asymmetric. This asymmetry is brilliantly exposed by Hegel when he says : « ...in the Ancient world only one man is free ; in the Helenic culture a few men are free ; but we know that all men are free » (Hegel, 55).

The internal fight for liberation from slavery is rather similar to the process of international struggle for surmounting the old asymmetries of the past. But it is more complicated, because the internal oligarchies are almost always confabulated with the dominant groups of the imperialistic countries[9]. However there is a similar process of liberation. The internal social and political dominance of small groups is overcome, through a process that can be very painful and, not a few times, tragically blooded.

---

[9] Perhaps some persons could think that I am old-fashioned because I employ the words « oligarchy » and « imperialism ». I would be old-fashioned if I thought that the cause of every evil in the world was due to the Washington government, and that communism will succeed in the whole world in a few years. But the fact that the Soviet Empire has collapsed and that the *ghost* of communism does not threat Europe any more, as it dit in the times of Marx must not be a pretext to lie about human history. Imperialism has existed and still exists, not only in international economy, but also in the cultural world. Of course nobody thinks nowadays that the Pentagon and the AIC are the source of all the evils. And nobody is silly enough to think that capitalism is the most cruel imaginable economic system. Thanks, in some way, to Marx and the Marxists, modern capitalism has very much evolved concerning the rights of the workers, and it is now practically bearable. But this does not mean that capitalism will reach fully social humanization. At least, for the time being. However, the world economy is changing very rapidly due to technological progress and the world expansion of informatics. I think that, in not a very distant future, capitalism will be superseded by some kind of welfare scientific-technological economy.

9.3.2 Cultural assymetry

Cultural asymmetry is tightly related to political asymmetry, but it is much more subtle and, many times, it is difficult to be aware of its existence. There is a *culture of domination* exerted through the cultural products of industrial most advanced countries. For instance, the famous cartoons that are read by children all around the world, contain an implicit message on the superiority of a nation over all the rest ; or of a « superior race » compared to the « inferior ones ». In this respect, the influence of television, radio and cinema is quite negative on non-learned groups. And now this influence is being rapidly augmenting through informatics.

As it frequently happens, once this subtle cultural domination is made conscious, generally by the intellectuals of the developing nations, there are strong contrary reactions. But these reactions are accompanied by a great danger. Some politically minded groups of weak nations fall very easily into fundamentalism, and this possibility is very dangerous indeed. The fundamentalist leaders have an insatiable thirst for power and they try to impose a full blooded terrorism against their opponents. It is worthwhile to relate a fact that happened some decades ago in a developing country. The leftist government tried to forbid the teaching of english in secondary school because it was the language of imperialism. Fortunately the *intelligentzia* reaction against this absurdity was so great, that the government had to drop it.

But the gravest problem of cultural asymmetry is the relation of culture with social situation. Generally, learning is the privilege of *élites*. Most members of the *élites* have noble ideas, and many of them strongly fight in favour of the poor classes. But in every country and, particularly, in the developing ones, this asymmetry imposes an implacable predetermination since the very moment of birth. The children of learned families are destined to command ; the children, of the unlearned families are doomed to poverty and obedience. There are exceptional cases in which a member of the poorest classes is able to conquest fame, honours and power. But the rule is that the son of a peasant shall be a peasant, and the child of a member of the *élite* will be a member of the *élite*.

The members of the *élites*, in many cases, try to change this situation. But then, they must face a vicious circle. On the one hand, to overcome poverty there must be a great deal of resources. On the other hand, because the country is poor, these resources do not exist. To really raise the life-standard of the poor, foreign investment is needed, and foreign investment means the imposition of economic standards by the powerful multinational groups. This situation reinforces the cultural dependence of the poor. There is a possibility to overcome this circular scourge. But it can only be done through the terrible sacrifice of one or two generations of peasants, workers and low middle class employees. This means that we are still very far from the realisation of the rational life ideal in the « so called (almost always with non-dissimulated contempt) the Third World ». But this is tantamount to the prevailing of arbitrariness and asymmetry in the national organisation of societies.

10. THE ROAD OF HISTORY

The previous considerations show, so I think, that the situation of nations concerning arbitrariness and asymmetry, is about the same within then and among them. As I have said (section 5), since the ending of World War II, in the international scenario, war was absolutely condemned. But behind the stage machinery it was wickedly incited and oriented by the lust of power and the greed of money. Arbitrariness and asymmetry was the rule. It might be thought that this assessment is only based on prejudice. But this is false. Because one of the causes of the present deplorable international situation is that the ideal of rational life is differently perceived within a single nation and within the international community. If one makes a serious inquiry within a country asking if the autotelic principle has been and is still being violated we can, for sure, predict that the majority of the answers will be, « yes it is ».

However, if the pool is made on international relations, the answers shall be quite different. A small minority will answer negatively ; but a great majority will say that the international policy of their country is approvable. In the first case the persons that answered « no » will automatically arouse the popular wrath, and be treated as traitors to their country. The cause of this situation is that most people believe the internal national policy and the international one, to be different. The belief of the population about the ethics concerning international relations is always the same : my fatherland is always right. This *credo* has imposed, through many centuries, respect and admiration for warlike personalities that plundered pacific countries and enslaved innocent people. The teaching of history in elementary and high school induces enthusiasm for the aggressive warfare and courageous warriors. It is evident that, as far as this difference between internal and external ethics, is maintained, it will be utterly impossible to realise justice among nations. The whole matter boils down to the following simple prescription : *the ethics of nations and of persons must be the same.*

11. CHALLENGING SISYPHUS

We have now a definite panorama concerning the political problems that characterise the final decades of the XXth century. I have proposed the criterion of the rational life ideal to cope with them. The ideal to transform history through the rationality of international and national behaviour, aims at the forging of a society in which arbitrariness and asymmetry have disappeared. It would be naive to imagine that this aim is fully attainable. But nobody can deny that in spite of all the horrors that our kind has suffered through the ages, humanity is advancing toward the supreme historical end : the forgery of a non-arbitrary and symmetric society, that is, the creation of a fully humanised world.

What can philosophers afford to this grandiose project ? To clearly see the possibilities of philosophy we must be aware that a philosopher is a person engaged with rationality[10]. Even the ivory-tower philosophers are engaged with reason, because they have chosen a life devoted to rational knowledge. Because of this engagement they are contributing to increase the richness of human knowledge. And what they find can be of the utmost utility to the achievement of rationality in history.

Anyway, it is obvious that even ivory-tower philosophers do not approve social arbitrariness and asymmetry. And it is certain that a majority of philosophers is explicitly opposed to them. But, to oppose them, a philosopher has only a sole possibility : *denouncement.* When a philosopher has decided to denounce exploitation and tyranny, he has powerful means to do it. Because his language is, generally, precise, and his rhetoric is rich in resources[11]. He must denounce irrational procedures

---

[10] The word « philosopher » has several meanings. But we can include all of them in the following dichotomy : philosopher *lato sensu* and philosopher *stricto sensu*. In the first sense Nietzsche is a philosopher ; in the second he is not. An irrationalist cannot be strictly considered as a philosopher because philosophy, above all things, is the endeavour to cope rationally with theoretical and practical problems. Relativists, empiricists, historicists are philosophers because, although they cannot be fully considered as rationalists, they utilise reason to try to demonstrate that what they say about knowledge is true.

[11] There is another possibility : participation in politics. But it is very difficult for a philosopher to find a favorable occasion to participate. However, this difficulty is not absolute. There are cases in which a philosopher has become involved in politics through ideological commitment, for instance, being the ideologist of a great political party. And there are some examples of philosophers that have actively participated in political action. For instance, myself. But this is a matter to be exposed in other kind of text. Anyhow, I think it is worthwhile to say a few words about the experience I had as an active member in Accion Popular, at the time the most important political party of Peru. I began by being the ideologist of the party and afterward I became the Vice Secretary of Politics. The most salient experiences I had during my active political life were the following. Politics is nowadays the same as in Ancient times. The impact of rhetoric is enormous, just like it

through the press, in lectures, in papers and in books. But to denounce irrationality one must have courage. In some situations he must be a hero. And to ask philosophers to be heroes exceeds the limits of human condition. But be it as may, the philosopher must be permanently alert regarding all the opportunities that offer him/her the possibility of denouncing and criticising the unjust procedures of the society he lives in.

One of the gravest problems philosophers have to face is the way history advances. As I have said, this advance is not continuous. It is like a zig-zag in which progress and recoil are always present. And this situation may induce him to pessimism. Discouragement is, perhaps, the greatest danger for a philosopher. Because if he loses his will to fight for justice, he will abandon the ideal of rational life. When a philosopher contemplates the atrocities that are happening every day in so many regions of the world, he feels himself like a Sisyphus. It is as if he were carrying an enormous and heavy rock to the summit of a mountain and, just when he had almost succeeded, the rock began to tumble down in spite of all the efforts to maintain his grip.

This situation has been dramatic since the Universal Declaration of Human Rights. When the Chart was proclaimed most philosophers and, with them, millions of persons, felt that the summit of the mountain was, at last, conquered. But, with awful promptitude, the Declaration was violated in every respect, as well in the international as in the national internal fronts. Is it still possible to ask a philosopher to continue his efforts in spite of the horrible tragedies that seem to have no end ? I think it not only to be possible but that, in the long run his efforts, against all immediate appearances, will have a perceptible effect. Because a philosopher is not a Sisyphus. He is not doomed to eternal failure by the jealousy of Zeus. A philosopher is a free man. And this means that, happens what happens, he can surmount discouragement. Perhaps the rock he is trying to drive to the summit will never reach the goal, but every time he will be able to lift it a little further.

---

was in the Greek Agora or in the Roman Forum. Philosophical wisdom found in many texts of the major philosophers through history, allows to know in advance the ways politicians follow to overwhelm their rivals. With a few honourable exceptions, their are able to lie, to slander and, if this is not sufficient, they resort to blackmail and even to murder. Of course I knew that most politicians believe that they can hold power for ever. And the more mediocre a politic is, the more he will abide to power. I knew all this *a priori*, but I had learnt it *a posteriori* reading philosophical and historical texts.

BIBLIOGRAPHY

DUVERGER Maurice, *Sociologie de la Politique*. Paris, Presses Universitaires de France, 1973.

EASTON David, *Enfoques sobre Teoria Politica*. Buenos Aires, Amorrortu, 1969.

HEGEL G.W.F., *Vorlesungen über die Philosophie der Weltgeschichte*. Hamburg, Felix Meiner Verlag, 1955.

LUCAS VERDU, *Principios de Ciencia Politica*. Madrid, Editorial Tecnos, 1979.

DAHL Robert, *The Behavioral Approach in Political Science*. New York, Hilt Rinehardt and Winston, 1969.

DEUTSCH Karl W., *Los Nervios del Gobierno*. Buenos Aires, Paidos, 1969.

MIRÓ QUESADA CANTUARIAS Francisco, *Hombre, Naturaleza, Historia*; Actas del Congreso Internacional Extradinario de Cordoba. Universidad Nacional de Cordoba, 1987.

MIRÓ QUESADA RADA Francisco, *Introduccion a la Ciencia Politica*. Lima, Cultural Cuzco S.A., 1994.

PARSONS Talcott, *Sociological Theory and Modern Society*. New York / London, The Free Press / Collier-MacMilland Limited, 1967.

PERROUX François, *La Coexistencia Pacifica*. México, Fondo de Cultura Economica, 1960.

– « *Independence* » *de la Nation*. Paris, Aubier-Montaigne, 1969.

TOFFLER Alvin, *El Cambio del Poder*. Barcelona, Editorial Plaza y Janes, 1990.

VAN DYKE Vernon, *Ciencia Politica : Un Analisis Filosofico*. Madrid, Editorial Tecnos, 1962.

WEBER Max, *From Max Weber, Essays in Sociology*. New York, A Galaxy Book / Oxford University Press, 1958.

YAKOVLEV A., *Conocimientos Politicos Basicos*. Moscou, Editorial Progreso, 1975.

REASONABILITY AND THE COSMOPOLITAN IMAGINATION : ARENDT, KORSGAARD AND RAWLS

David M. RASMUSSEN
Boston College, Boston

The idea of reasonability can be expanded to include a kind of reciprocity between western and non-Western peoples which is asymmetrical. One implication of this thesis is to move us beyond a conception of reciprocity based on a conception of autonomy. Fundamental to this transformation is a reconsideration of Kant's political theory based not on a notion of autonomy but on a notion of sociability, which will in turn allow a reconfiguration of the notion of normativity. My hypothesis is that a consideration of the *Critique of Judgment* will enable us to get beyond a theory of identity implicit in a reflection on reasonability from the point of view of autonomy. Reasonability, which has its origin in contemporary discussion in W.M. Sibley's 1953 article, entitled « The Rational vs. the Reasonable » and reworked by John Rawls in his post-1979 writings supports an ever-widening understanding of normative relations, which are not necessarily reciprocal. Hence, reasonability as associated with public reason opens up a kind of intersubjective understanding from which it is possible to widen the perimeters of social co-operation. From this position the notion of normativity must itself be worked out.

I have argued elsewhere that even though Rawls in his *Law of Peoples* rejects a cosmopolitan interpretation of his work, his argument requires a weak cosmopolitanism[1]. Kant, in his famous fifth thesis of « Idea for a Universal History with a Cosmopolitan Intent » argues, « the greatest problem for the human species, whose solution nature compels it to seek, is to achieve a universal **civil society** administered in accord with

[1] Rasmussen, David M. « Beyond Liberalism : Toleration and the Global Society in Rawls' *Law of Peoples* » in *Pluralism and the Pragmatic Turn* : ed. William Rehg and James Bohman. Cambridge: MIT Press, 2001, p. 389-412.

P. Kemp (ed.), World and Worldhood, 187-205.

right »[2]. Although I am in agreement with the overall intention of this claim, in my view it must be regarded as not sufficiently differentiated. A liberal society cannot simply address other non-liberal societies with the immediate assumption that they should mirror its identity. This modification, however, does not mean that we cannot retain the idea of cosmopolitanism. Why ? Because cosmopolitanism is not simply a concept. Rather it is an aesthetic idea, which gives purpose to history from the perspective of the human agent or subject. Fundamentally, the term ‘ cosmopolitan ’ owes as much to sensibility as it does to cognition. Whatever is normative about global justice within a cosmopolitan framework will be built up over time through the interpretations and reconstructions liberal and non-liberal societies mutually accept.

Public reason is often conceived as an exclusively Western concept, the brainchild of liberal political thought, anchored in a commitment to cosmopolitanism that was born in the enlightenment and dismissive of non-Western points of view. Perhaps political thinkers in the West have been guilty of simply assuming the post-enlightenment stance of public reason with its implicit commitment to impartiality, agnosticism regarding religious values and the innate superiority of Western ideas. However, I believe public reason and global justice are not mutually contradictory. Indeed, as Rawls states in the first version of his *Law of Peoples*, « whenever the scope of toleration is extended : the criteria of reasonableness are relaxed »[3].Today as we experience a certain crisis of global culture one wonders if public reason can bridge the gap between Western and non-Western points of view. From our current perspective one must acknowledge that Rawls was prophetic in his *Law of Peoples,* anticipating the extent to which it has become necessary to accommodate the views of non-Western, non-liberal societies. To some it may appear that he went too far, simply abandoning or attempting to abandon a cosmopolitan point of view. In light of recent events we can see that his subordination of reason to reasonability, which emerged as a principle in his thought in 1979-80 and continued to guide his reflections thereafter, provided the possibility of an accommodation to non-liberal, non-western perspectives which we now most definitely need.

---

[2] Kant, Immanuel. *Perpetual Peace and Other Essays.* (Trans. Ted Humphrey) Indianapolis : Hackett Publishing Co., 1983, p. 3.

[3] Rawls, John. *Collected Papers.* (Ed. Samuel Freeman) Cambridge : Harvard University Press, 1999, p 561.

In this essay I will consider the notion of reasonability and its normative implications as fundamental for the construction of a notion of public reason, which is cosmopolitan in scope. First, I want to look at the argument for reasonability. Second, I want to consider the normative implications of reasonability for an understanding of public reason. Third, I will consider the post-autonomous implications of the notion of reasonability, i.e. the relationship between reasonability and judgment. Finally, I will return to the cosmopolitan implications of the notion of reasonability.

I. REASONABILITY AND REASON

The distinction between reason and reasonability enters into contemporary discussion in the 1953 article by W.M. Sibley entitled « The Rational and the Reasonable ». Sibley there establishes a number of now familiar distinctions between the rational and the reasonable arguing that the term « reasonable » covers areas not intended by the term « rational » when considered in relation to human conduct. In his language, « with regard to the *ends* I propose to myself [the use of the term, « rational »] entails (a) that I have an informed awareness of the nature of the ends I am proposing to achieve, including in this awareness a realization of their significance as they affect the other ends, not only myself, but also others affected by my actions ; and (b) that in case of a conflict between two of my proposed ends, I select that end which I really prefer, i.e., that end which, after informed and careful reflection, taking into account my own experience and what I know of the experience of others, I judge to be of more value than its competitor »[4]. As for the means for attaining those ends Sibley believes it is enough that they be the best[5]. To the extent that the will is involved, Sibley states that one should act in accord with decisions reached through the « process of reflection » without the influence of one's « emotion ».

In this view, rational conduct takes into account the interests of others only to the extent that they effect the rational actor's own interests. However, according to Sibley there is another way in which the interests of others may be taken into account, namely, taking account of the

---

[4] Sibley, W. M. « The Rational Verses the Reasonable ». *The Philosophical Review,* Volume 62, Issue 4, 1953, pp. 555-6.
[5] This echoes the classical understand of reason in the context of teleology.

interests of others as a « disinterested impartial spectator might do ». In this case the particular interest of the other is put on a level with one's « own » interest. For Sibley the latter « requires a positively sympathetic disposition toward others »[6] as well as « a preparedness to be genuinely concerned with ' their ' interests per se ». Sibley characterizes this disposition as a « *moral* virtue » which requires the special designation « reasonable ».

In this view not all reasonable conduct is directed toward the other. There are occasions when reasonable and rational can be conceived as almost identical. Sibley uses the example of making a reasonable investment, which turned out badly. Rational self-interest prevails in this case even though we attribute reasonableness to the action. A moral judgment involves a consideration which takes into account the « other person's point of view and, in so doing, takes that person's point of view into account so that one could in fact be ' influenced ' by it ». Sibley concludes that « reasonableness thus requires impartiality, ' objectivity ' ; it expresses itself in the notion of equity »[7].

What Sibley does not capture, and what has dominated the discourse on reasonability, is its normative or obligatory character. Reasonability carries with it the notion that the reasonable point of view entails an obligation to consider the judgments of others in a context which is free from instrumental, self-interested reason. So when reasonability provides the basis for the normative claims of public reason in a political sense and for the normative content of moral theory, it is on the justification of our obligation that attention is focused. The authority for obligation is controversial in the sense that not everyone would agree on the same kind of authority. Certainly when one turns to a global context it would be difficult to make the claim that there is one universal authority to which obligation should point.

John Rawls entertains a much more complex notion of reasonability than does Sibley. I have argued elsewhere that Rawls' distinction between the rational and reasonable should be seen as strengthening rather than weakening his position[8]. Essentially it does this by shifting the normative status of the theory of justice beyond the

---

[6] *Ibid.*, p. 557.
[7] *Ibid.*, p. 558.
[8] Rasmussen, David. « Reasonability vs. Reason : Reflections on the Reasonability of Public Reason. » Scheduled to appear in the proceedings of the Italian-American Conference on Philosophy.

narrow confines of a theory of autonomy. Whether or not this is a substantial move beyond the claims of Kant's moral theory or not is a matter subject to interpretation[9]. What is clear is that Rawls eventually re-interprets Kant's moral theory in such a way that the notion of autonomy is broadened to have a social focus. It is in this sense that the intersubjectivist hypothesis implicit in Arendt's claim about Kant's politics conceived in relationship to the idea of sociability (to which I will turn in a later section of this essay) could be correlative with Rawls' move to public reason. I will argue, contrary to much of the interpretative literature on the development of Rawls' work, that this move to an intersubjectivist orientation and therefore the move to public reason was implicit from the outset. The first reason for making this claim is to be found in Rawls' early critique of utilitarianism with its assumption about rational egoism and efficiency. Rawls acknowledges at the end of his 1958 article, « Justice as Fairness », that his « modification of the utilitarian principle » was « slight » even though it would « have a different conception of justice standing behind it »[10]. When he tried to clarify this ' slight ' difference he indicated how his conception of justice required or involved the « mutual acceptance... of the principles on which the practice is founded »[11]. This mutual acceptance of the principles involved would eventually drive him beyond a conception of reason based on the egoistic interests of the actors alone even though the conception of reason based on self-interest would dominate. Hence, it is my view that the thesis that the original position was limited by its employment of a framework taken from rational choice theory misses the point that the critique of utilitarianism was initially intended to provide a framework for mutual acceptance[12] .

The second reason for making the claim that the intersubjectivist move was implicit in Rawls' theory from the outset is to be found in the 1963 article, « The Sense of Justice » which begins by affirming

[9] What I mean by this is that by translating the word, *Vernünftig*, as meaning both reason and reasonable Rawls is able to develop a theory of intersubjectivity within Kant's moral theory. Hence, he is able to move beyond a theory of autonomy within a theory of autonomy.

[10] Rawls, John. *Collected Papers*. Cambridge : Harvard University Press, 1999, p. 71.

[11] *Ibid.*

[12] Critics of Rawls have argued that he was hampered by his use of rational choice theory. In a sense they are correct. But the argument is limited. This criticism misses the point that the role of mutual acceptance would drive the theory of justice beyond the autonomous foundations of rational choice theory.

Rousseau's assertion that « the sense of justice is no mere moral conception formed by the understanding alone, but a true sentiment of the heart enlightened by reason, the natural outcome of our primitive affections »[13]. No doubt Rawls came to see that on of the implications of justice as fairness was to be an underlying theory of obligation. Hence, Rawls speculates that if one is to conduct oneself in accord with certain principles as is required in the original position it is because one possesses a sense of justice conceived as a capability. Of course, this is a moral notion that conceives of obligation as a constraint on behavior. Rawls' way of putting this is itself constrained by the language of self-interest. « The concept of morality, when imposed upon rational and self-interested persons, gives rise to certain definite constraint »[14]. However, in order to fulfill the obligations to others under justice, the assumption is made that one must be able to have the capacity for justice ; « the capacity for a sense of justice, is then, necessary and sufficient for the duty of justice to be owed to a person »[15]. This obligation is based on autonomy inasmuch as the duty of justice, in his view, is dependent on a person being « regarded as holding an initial position of equal liberty »[16]. It is the notion of autonomy that distinguishes Rawls' position from utilitarianism with its emphasis on pleasure and pain as sufficient justification for a theory of rights and obligations. Equally, although the idea of capacity has a certain Aristotelian ring to it, Rawls would differentiate himself from that view by pointing out that the underlying assumption of the classical view is a theory of inequality.

My argument is that the social, intersubjectivist point of view is already implicit in Rawls' earliest writings in his critique of utilitarianism and in his developing theory of obligation. But I would also argue that position lacked a social perspective, developed as it was on the basis of a narrowly focused Kantian idea of autonomy. To the extent that reasonableness was to be included in that early view, there is a certain truth in the critique, which Rawls later rejected, that the reasonable is derived from the rational. In any case, I contend that when Rawls moved to the distinction between the rational and the reasonable with the subordination of the rational to the reasonable, the normative status of his argument changes. Instead of conceiving morality as a mere constraining

---

[13] *Ibid.*, p. 96.
[14] *Ibid.*, p. 99.
[15] *Ibid.*, p. 115.
[16] *Ibid.*, p. 115-16.

of mutual self-interest, reasonability with its implicit obligation regarding the political interests of others became an independent normative criterion. It will follow that from the point of view of the theorist the idea of political obligation as well as its justification will predominate over moral obligation. As I suggested a moment ago, Rawls found a way to justify this shift in his thinking from within Kantian moral theory through his interpretation of *Vernünftig* in Kant's work as meaning both rational and reasonable[17]. However, although the two may have a common source the reasonable cannot be derived from the rational. It follows that the introduction of reasonability represents a new normative criterion. Rawls contends that « the reasonable is public in a way that the rational is not »[18]. He goes on to suggest that « it is by the reasonable that we enter as equals the public world of others and stand ready to propose, or to accept, as the case may be, fair terms of social cooperation with them »[19]. The reasonable then represents the introduction of a public, social and in this sense intersubjectivist criterion. The claim is that « insofar as we are reasonable, we are ready to work out the framework for the public social world, a framework it is reasonable to expect everyone to endorse and act on, provided others can be relied on to do the same »[20].

At the same time the introduction of reasonableness represents a pragmatic acknowledgement of the diversity of modern culture as well as its fundamental plurality. The consequence is the acknowledgement of reasonable disagreement. The sources or causes of reasonable disagreement are classified under the title, ' burdens of judgment '. I would argue that the introduction of judgment under the category of reasonability is not to be directly identified with Kant's use of the term in the *Critique of Judgment*. However, Rawls' use of the term ' judgment ' does have a certain family resemblance to that use in that it acknowledges particularity. Different forms of evidence, the relevance and weight of evidence, the vagueness of concepts and their interpretation, the effect of the particularity of experience, different normative considerations, conflicting values – all of these and perhaps other factors or sources affect our judgments and may be classified as part of the burdens of judgment. One can expect as a consequence, as Rawls acknowledges, that

---

[17] Rawls, John. *Lectures on the History of Moral Philosophy.* Cambridge : Harvard University Press, 2000, pp. 164-5.
[18] Rawls, John. *Political Liberalism.* New York : Columbia University Press, 1993, p. 53.
[19] *Ibid.*
[20] *Ibid.*, pp. 53-4.

universal agreement is an unrealistic expectation in a public political culture. He concludes the discussion by stating that « these burdens of judgment are of first significance for a democratic idea of toleration »[21].

The idea of reasonability is a consequence of Rawls' early attempt to work out a theory of obligation. Significantly, it goes beyond a Kantian notion of autonomy, not by abandoning it but by moving to a public context. In this sense, Rawls' idea of reasonability differs from that of Sibley by working out the social or political context in which it occurs. Rawls specifically declines to characterize reasonability as an epistemological procedure. His claim is that « it is part of the political ideal of democratic citizenship which includes the idea of public reason »[22]. The idea of reasonability under the burdens of judgment includes the idea of respect for the views of others which takes into account their reasonable comprehensive views. Equally, it establishes the expectation that others will present their views in a reasonable manner. The constraint on citizens is the givenness of political culture that provides the context for political interaction.

II. NORMATIVITY

We could begin by claiming that reason and reasonability have the same meaning and therefore rest on the same normative claim. Both reason and reasonability are based on obligation to others, but that obligation in each case is different. Following Sibley, reason does have an obligatory meaning in the sense that conduct is directed towards particular ends, and the choice of means to pursue those ends does take account of the interests of others. Reasonability entails a quite different kind of obligation. Reasonability acknowledges the interests of others in the sense that it can represent those interests with certain impartiality. By so doing, reasonability harbors the principle of equity because it involves a reflective representation of the interests of others on a par with one's own interests. Equity and impartiality can be said to be normative principles, because they represent a particular concept that regulates social interaction. They represent norms in the sense that they structure a particular obligation. I refer here to the origin of the term « norm », which can be traced back to the Latin *norma,* the term for a builder's

[21] *Ibid.*, p. 58.
[22] *Ibid.*, p. 62.

square which gave a certain « standard of correctness » prior to the actual construction of a building[23]. The standard or norm appropriate to the notion of reasonability is associated with equity, impartiality, willingness to take the position of the other, even fairness.

In her book, *The Sources of Normativity,* Christine Korsgaard presents a justificatory argument for the normative character of morality and/or ethics[24] .One of her major claims is that modernity makes morality normative. Her argument is that what was fundamental for the classical worldview, i.e., that the world had a purpose, is no longer the case with the development of the modern scientific perspective on the world[25]. When the modern scientific worldview took the idea of the purposiveness of the world away morality needed an independent justification. According to Korsgaard, with modernity four different justifications for morality emerge. Voluntarism begins with the assumption that since values are not to be found in the world they have to be imposed. In voluntarism, authored by the contractualists Pufendorf and Hobbes, it is assumed that morality originates through obligation that in their view can only be secured by a « godlike sovereign ». The point being that « normativity springs from the legislative will »[26]. As my earlier reflections on Hobbes suggest, his argument is a bit more complex than that inasmuch as his concern with « whose reason it is » makes the question of legislative authority more complicated[27]. However, for Korsgaard's purposes the generalization can stand. Normativity is derived from legislative authority that makes obligation possible.

Realism responds to voluntarism in this view by saying « normative claims are normative only if they are true »[28]. Korsgaard's claim about realism begins with reference to Clarke's critique of Hobbes. Clarke questioned why it was necessary to be forced to obey the law of

---

[23] See Dancy, Jonathan. *Normativity.* For a clarification of the origins of the term « norm » see Railton, Peter. « Normative Force and Normative Freedom : Hume and Kant, but not Hume *Versus* Kant ». In that volume, pp. 1-2.

[24] Korsgaard, Christine M. *The Sources of Normativity.* Cambridge : Cambridge University Press, 1996.

[25] No doubt she adopts Kant's view at this point, namely, that if we see nature as purposive it is because we superimpose a purpose onto nature.

[26] *Ibid.*, p. 19.

[27] In the essay entitled « Jurisprudence and Philosophy » I considered the complex argument regarding legislative authority which Hobbes displayed in his argument with Lord Coke.

[28] *Ibid.*

nature if nature « obligates » us anyway. The realist shifts the argument from the question of power and force to the problem of motivation. Why is it necessary to demonstrate a need for obligation when it is obvious that people do obligate themselves ? So the argument for the reduction of authority to power disintegrates in light of the fact that we choose to be obligated. From the realist point of view obligation is a fact.

Korsgaard associates Hume's critique of realism with a reflective endorsement model. This model originates from the fact of human sentiment. Korsgaard's example is Hume's view of murder. Murder is not disapproved of because it is wrong, it is wrong because it is disapproved of [29]. The argument that we have certain moral sentiments and through reflection we endorse those sentiments assumes that reflection on moral sentiment enhances that sentiment. It is Korsgaard's contention that the reflective endorsement approach to normativity is but a development of the realist's claim. The argument is that if obligation is a given as part of human nature as the realist would contend, reflective endorsement carries that normative claim to a higher level.

From Korsgaard's perspective, what we learn from all this is that the progressive development of normative justification of moral theory in modernity focuses on reflection. The fundamental problem then becomes the problem of the « authority » of reflection. If obligation is the product of our own nature, one formulation could be « autonomy is the source of obligation ». This is not an intersubjective normative claim in the sense that the formula « autonomy is the source of obligation » is derived from the categorical imperative. The latter is derived from acting on maxims and not from acting on obligations to the other. In Korsgaard's view the moral law is derived from the kingdom of ends which can be formulated as « we have obligations to humanity as such ». In other words, the original claim is derived from autonomy as such. The « autonomy is the source of obligation » is a reformulation of the categorical imperative, while the « obligation to humanity » formulation refers to the « kingdom of ends »[30].

[29] The point being that human beings establish and endorse moral standards, as opposed to those standards being given by nature.

[30] She states, « but it is true that the argument that shows that we are bound by the categorical imperative does not show that we are bound by the moral law. For that we need another step. The agent must think of *herself* as a Citizen of the Kingdom of Ends ». *Ibid.*, p.100.

The movement from voluntarism through realism through reflective endorsement to the « autonomy as the source of obligation » formula is a movement from external legislative authority to the realization that the source of legislative authority is personal identity. In Korsgaard's words, « so the reflective structure of human consciousness establishes a relation here, a relation which we have to ourselves, and it is a relation not to mere power but rather of *authority*. And *that* authority is the source of obligation »[31].

III REASONABILITY

Korsgaard provides a framework for understanding the sources of normativity. I will argue a parallel but slightly different case when we turn to the normative foundation for reasonability. We could understand reasonability under the category of the universal law associated with the process of universalization. However, if reasonability is to be understood as a political and not solely as a moral, idea then the kind of appropriation of and obligation to the other has to be presented in such a way that the other is not reduced to the mirror image of the self to which it is related. If we approach the question of the normativity of reasonability from the perspective of moral theory, we run the risk of a conceptual identity, which fails to take into account the other as other. The problem is that cosmopolitan orientations based on Kant's moral theory tend to construe fundamental ideas like impartiality, equity, and fairness, ideas fundamental to reasonability, on the basis of a certain conceptual identity. The end result is that global encounters between non-Western and Western points of view take place on a model in which the other is accounted for but only on the basis of a conceptual identity. I propose a slightly, although perhaps not a radically different, route.

What reasonability requires is a particular kind of attitude, which is sensitive to the other, an attitude which incorporates autonomy and yet goes beyond autonomy. Such an attitude is based on judgment because it must go beyond an orientation of the self to itself. In peculiar ways both Korsgaard and Rawls, who both trace their positions back to Kant, go beyond the Kantian notion of autonomy. Korsgaard goes beyond autonomy by affirming the kingdom of ends as a social or intersubjective category and by endorsing Wittgenstein's argument against private

[31] *Ibid.*, p. 104

language. Rawls goes beyond the Kantian notion of autonomy by subordinating reason to reasonability which he grounds in an interpretation of Kant's use of the term *Vernünftig*. A third way to make this move and still remain faithful to Kant would be to turn to Kant's notion of aesthetic judgment for an elaboration of reasonability. The claim is that the intersubjectivist understanding which reasonability presumes can be grounded in aesthetic judgment.

I find the argument worked out in Hannah Arendt's *Lectures on Kant's Political Philosophy*[32] suggestive for that purpose. Arendt developed two worthwhile theses regarding the interpretation of Kant's *Critique of Judgment*. Her first thesis is that aesthetic judgment rests on an assumption regarding human sociability and as such can be classified within the Kantian corpus as a political claim. Her second thesis implied by the first is that politics as such requires judgment for its justification. Unfortunately, at the heart of her analysis she allies the distinction between contemplation and action with Kant's distinction between taste and genius. That is a kind of third thesis, which cannot be supported by Kant's text [33].

With regard to the first thesis according to Arendt, when placed under the topic « affairs of men » Kant's overall work as a philosopher can be divided into three topics, i.e., « human species », « man » and « men ». Under the first, human species, subtopics like mankind, the human being as part of nature, and human being as subject of history are considered under the idea of « end ». If we look for the work in which the human species is considered in this way we would look, according to Arendt, at the « Critique of Teleological Judgment », i.e., the second part of the *Critique of Judgment*. Under the topic « man » we would consider a « reasonable being » who is « subject to the practical laws he gives

[32] Arendt, Hannah. *Lectures on Kant's Political Philosophy* (edited with an Interpretative Essay by Ronald Beiner). Chicago : The University of Chicago Press, 1982. For a comprehensive analysis of Kant's theory of judgment as applied to contemporary political philosophy see Ferrara, Alessandro. *Justice and Judgment*. London : Sage, 1999.

[33] A particularly precise critique of Arendt's spectator/actor distinction is to be found in Paul Ricœur's *The Just*. Translated by David Pellauer. Chicago : The University of Chicago Press, 2000, pp. 94-108. Although Ricœur admires Arendt's reconstruction he points out that political judgment requires the actor as well as the spectator in the sense that without the action of the revolutionary there would be nothing to judge. Equally, he regards the emphasis on contemplation to eliminate the prophetic role of aesthetic judgment. The example of war makes the point. It is only tolerable from the point of view of the spectator, but from the point of view of moral judgment it is absolutely intolerable.

himself, » who is both « autonomous » and an « end in himself, » who « belongs to the realm of intelligible beings ». Kant covers these topics in the *Critique of Pure Reason* and *the Critique of Practical Reason.* Under the topic « men », Kant considers human beings as « earthbound creatures » who live in « communities » and are « endowed with common sense » and « *sensus communis,* » they have a « community of sense » and they are « not autonomous ». They « need each other's company » and they enjoy the « freedom of the pen ». So interpreted Arendt can support the thesis that politics conceived under the category of sociability is the topic of the first part of the *Critique of Judgment*[34] .

What seems to lie behind this way of considering both Kant's overall work and the distinctive claim about the critique of aesthetic judgment is that only under that topic does Kant consider the « sociability of man », in other words, human beings in their common association. What do we presume when we, as human beings, live together ? What is normative for our association ? These are political questions according to the classical meaning of the term « politics ». Hence, it is not entirely inappropriate to consider aspects of the Third Critique as contributing to an understanding of politics. Arendt believes that Kant was never able to write a politics directly because of the three large questions he asked : the question of the human being as a political animal wasn't one of them. The questions, « what can I know ? » « what ought I to do ? » and « what may I hope ? » do not address what the human being is as a « political animal »[35]. Arendt believes aesthetic judgment addresses the political question.

If aesthetic judgement is based on human sociality it can be appropriated for politics. Politics in this view is based on the movement from the particular to the universal will require a particular kind of justification which can be based on the elaboration of judgement. The second thesis is more complex than the first. In judgment the imagination overcomes the particularity of taste through the « operation of reflection », which overcomes particularity by « removing the object ». And when this is done one has « established the conditions of *impartiality* »[36] and those very conditions result in sociability. In order to support this claim she quotes Kant's *Critique* from both #41 « the beautiful interests [us] only [when we are] in *society* » and #45 [Man] is

---

[34] *Ibid.*, pp. 26-27.
[35] *Ibid.*, pp. 19-20.
[36] *Ibid.*, p. 67.

not contented with an object if he cannot feel satisfaction in it in common with others »[37]. The result of this is that « in taste egoism is overcome ». Because of the impartiality implicit in this process we come to understand that « we must overcome our special subjective conditions for the sake of others »[38]. For Arendt this means « the nonsubjective element in the nonobjective sense is intersubjectivity »[39]. It would follow that that tradition which found fault with Kant for the reliance on autonomy alone, a critique that may or may not have been valid, could have looked to his aesthetic writings for a more balanced view. As Arendt ingeniously observes, « you must be alone in order to think : you need company to enjoy a meal »[40].

One moves in the *Critique of Judgment* from the particularity of taste to the generality of communication by the activity of judgement. In judgment the « operation of the imagination » prepares the object for the operation of reflection. In Arendt's view it is this second operation, the operation of reflection, which « is the actual activity of judging something »[41]. Again following Arendt, « the criterion then is communicability and the standard of deciding is common sense »[42]. This leads to a discussion of #39-40 in the *Critique of Judgment,* where Kant transforms the idea of common sense into *sensus communis.* In #39 Kant claims that judgments of pleasure which involve the perception of the object by means of the imagination must be exercised « on behalf of the commonest experience ». In #40 he changes the term indicating that common sense in the elevated sense of *sensus communis* is that capability by which men are to be distinguished from animals. Arendt claims, « it is the very humanity of man that is manifest in this sense »[43]. So common sense becomes a kind of « community sense ». Hence the claim : « this *sensus communis* is what judgment appeals to in everyone, and it is this possible appeal that gives judgments their special validity »[44].

In the end Arendt sees aesthetic judgment with its commitment to an enlarged mentality, the ability to think from the standpoint of another,

---

[37] *Ibid.*
[38] *Ibid.*
[39] *Ibid.*
[40] *Ibid.*
[41] *Ibid.*, p. 68.
[42] *Ibid.*, p. 69.
[43] *Ibid.*, p. 70.
[44] *Ibid.*, p. 72.

as an analogue to Kant's late cosmopolitanism. This is important because it links Kant's aesthetics with Kant's late in life political observations. *Sensus communis,* then is a kind of cosmopolitan notion which would tie in with her view of the political subject, the world observer who understands the original compact with humankind as a whole or humankind as a totality. In politics then, given this foundation in aesthetics, it would be the particular that would make the case for the universal. « This exemplar is and remains a particular that in its very particularity reveals the generality that otherwise could not be defined »[45].

The claim that needs to be highlighted here is that, at least for the moment, we can say that politics works by examples, reason works through the concept or the schema. What Arendt wants to make is the following generalization : « The same faculty, imagination, which provides schemata for cognition, provides *examples* for judgment »[46]. This claim is important for the normative claim I wish to make about reasonability. In Arendt's view the imagination produces a synthesis which amounts to providing an « image for a concept » which we call schema. Hence, « what makes particulars *communicable* is (a) that in perceiving a particular we have in the back of our mind a « schema » whose « shape » is characteristic of many particulars *and* (b) that this schematic shape is in the minds of many different people »[47]. This leads to the distinction between determinate and reflective judgment. Determinate judgments subsume the particular under the general rule while reflective judgments derive the rule from the particular with the result that reflective judgment requires the attitude of reasonability.

The main point of this discussion is that in the *Critique of Judgment* the « analogy » to the schema is the example. It is not the case that the quest for validity is given up ; rather one finds validity in the example. The formula for this claim is « exemplary validity ». Arendt's concludes that in politics it is not the schema but the example that plays the leading role. « Most concepts in the historical and political sciences are of this restricted nature ; they have their origin in some particular historical incident, and then to make it « exemplary » – to see in the particular what is valid for more than one case » [48].

---

[45] *Ibid.*, p. 77.
[46] *Ibid.*, p. 80.
[47] *Ibid.*, p. 83.
[48] *Ibid.*, p. 85.

After a consideration of Arendt's reading of Kant's political philosophy the sources of normativity can be broadened to include a kind of reasonability associated with procedure outlined in aesthetic judgment. Here it is possible to stress Arendt's thesis that Kant's unwritten political philosophy stems from the idea of sociability. If one were to make this claim in relationship to Kant's moral philosophy it would suggest that the idea of sociability moves us beyond the idea of autonomy. However, following Korsgaard in her reflections on the ' kingdom of ends ' sociability is not a replacement for autonomy, it provides a way of moving beyond autonomy. Technically, the idea of sociability involves a fundamentally intersubjective claim which requires a kind of justification which includes, as Sibley pointed out in his discussion of reasonability, a particular kind of sensibility regarding other persons. If we are to follow the procedural model given in Kant's writings on aesthetics the phenomenon to be justified is particularity. Of course the particular does not remain particular, rather it must be justified beyond its status as particularity. If we follow the procedure of aesthetic judgment it is clear that the imagination through reflection transforms the particular into something universal. This transformation is attributable to *Urteilskraft,* the power of judgment which raises the particular to universality associated with exemplary validity. The peculiar requirement of judgment is a taking of the other into account through the process of postulating a kind of sensibility which is cognizant of others. To follow Sibley again, a certain principle of equity is required which involves taking into consideration the position of the other. At this point one could invoke Kant's idea of *sensus communis* which provides a substitute for the notion of universality derived from a logical procedure. It is in this sense that one can affirm Arendt's contention that the idea of sociability provides the basis for the normative justification of politics.

IV. REASONABILITY AND COSMOPOLITANISM

I have suggested that the cosmopolitan idea or the idea of cosmopolitanism is an aesthetic idea which can have either a strong or a weak interpretation. The temptation would be to associate the cosmopolitan idea with teleological judgment and not with aesthetic judgment. No doubt Arendt's attempt to ally politics solely with aesthetic judgment and the spectator robs any notion of justice of both its prophetic and historical role. She wants to dissociate politics from a strong philosophy of history of an Hegelian character. However, a careful reading of Kant might demonstrate that there is a certain truth to her claim about the link between politics and judgment.

The idea of reasonability requires judgment because it is necessary to consider the point of view of others and attribute reasonability to their views. For Kant, the rudimentary foundation for such attribution is found in the mere communicability of sensation. We simply presuppose on the most mundane level the universality of sensation, i.e. « that everyone's sense is like our own »[49]. However, this is not necessarily the case unless we take the next step adding to sensation a certain pleasure. So the question is, how do we experience sensation or pleasure ? This is the original complex question Kant wants to ask. In the most minimal language, for us to experience a sensation as pleasure a « procedure of judgment » is carried out which involves the imagination and the understanding. And since this condition is merely « subjective » it must be the same in everyone and therefore universally communicable. This ' original ' way of formulating the idea of the universal communicability of sensation is but a vulgar formulation in Kant's view. It is as if we derive the idea of common sense from sensation when in fact we don't. What we presuppose as universal, i.e. as attributable to everyone, is judgment. Hence, Kant would reformulate the common sense idea of the universal communicability of sensation according to the formula : « Taste as a kind *of Sensus Communis* »[50]. *Sensus Communis* is taken to be a sense shared by everyone defined as simply a universal power to judge that exists in everyone. Kant reminds us that in this comparison with everyone we take into account « human reason in general » so as to escape the illusion of deriving our claim from a merely private contention. Indeed, this is a transcendental justification in which we compare our judgments not with the actual but with the « possible judgments »[51] of others. In so doing, and this is Kant's crucial justificatory claim, we « put ourselves in the position of everyone else »[52] by abstracting from the possible contingencies of our own judgments.

The next step in the argument is not, as some seem to think, to ally *sensus communis* with « common human understanding » but to

[49] Kant, Immanuel. *Critique of Judgment.* (Trans. Werner S. Pluhar) Indianopolis :Hackett Publishing Co., 1987, p. 157.
[50] *Ibid.*, p. 159.
[51] *Ibid.*, p.160.
[52] *Ibid.*, p.160.

characterize the « aesthetic power of judgment » as a « shared sense » as opposed to an « intellectual » one.[53] Indeed, there is no conceptual justification of an aesthetic judgment of taste. Hence Kant's conclusion : « We could even define taste as the ability to judge something that makes feeling in a given presentation *universally communicable* without mediation by a concept »[54].

The latter claim is important for the attempt to rest a theory of politics on aesthetic judgment because the mode of justification will have to follow the formula « justification without a concept ». This justification can be developed in two ways. One is the procedural justification associated with a « shared sense » included in the procedure of judgment and the attribution of reasonability to the other. The other justification is associated with how one can exhibit the phenomenon which gives rise to the judgment of taste. Clearly, Kant has to answer the question « what is it that requires taste ? » His answer is the aesthetic idea, the « presentation of the imagination which prompts much thought, but to which no determinate thought, i.e. no [determinate] *concept,* can be adequate, so that no language can express it completely and allow us to grasp it »[55].

There is evidence to suggest that the model for the cosmopolitan idea is the aesthetic idea. In the ninth thesis of Kant's *Idea of a Universal History with a Cosmopolitan Intent,* Kant endorses such a view first by characterizing his own effort as something closer to aesthetics than science. He states : « It is, indeed, strange and for all appearances absurd scheme to write a history based on an idea of how the course of the world must go if it is to approach a certain rational goal : it seems that such an attitude can only result in a *romance* »[56]. To be sure the model for the cosmopolitan ideal is not science. Here Kant acknowledges that the technique for exhibiting the cosmopolitan idea is more like writing a novel than recording history. My assumption is that he intends the cosmopolitan ideal or idea to be more like an aesthetic idea.

> « If one may nevertheless assume that nature does not proceed without a plan and a final objective, even in the play of human freedom, this idea can still be useful and while we are too short

---

[53] *Ibid.*, p.162.
[54] *Ibid.*
[55] *Ibid.*, p.182.
[56] Kant. *Perpetual Peace and Other Essays*, p. 38.

> sighted to penetrate the hidden mechanisms of her workings, this idea may still serve as a guiding thread for presenting an otherwise planless *aggregate* of human actions as a system »[57].

Kant then leads to the following conclusion, namely that the cosmopolitan idea makes its normative claim not on the basis of a concept in the ordinary sense of the term. Rather the conception of sociability implicit in that idea need not relate self and other in terms of a conceptual identity. Hence, the cosmopolitan idea exhibits sufficient space to enable a notion of global justice, which moves beyond a strict reduction to identity.

---

[57] *Ibid.*

INDEX

MIX
Papier aus verantwortungsvollen Quellen
Paper from responsible sources
FSC® C105338

If you have any concerns about our products,
you can contact us on
ProductSafety@springernature.com

In case Publisher is established outside the EU,
the EU authorized representative is:
Springer Nature Customer Service Center GmbH
Europaplatz 3, 69115 Heidelberg, Germany

Printed by Libri Plureos GmbH
in Hamburg, Germany